In Memory of

Peter W. Souchock

Presented to the
Adrian Public Library

by

Staff of the

Adrian Public Library

GOLF
by the Numbers

ROLAND MINTON

The Johns Hopkins University Press
BALTIMORE

The Johns Hopkins University Press
2715 North Charles Street
Baltimore, Maryland 21218-4363
www.press.jhu.edu

ISBN-13: 978-1-4214-0315-1 (hardcover)
ISBN-10: 1-4214-0315-3 (hardcover)

Library of Congress Control Number: 2011929971

A catalog record for this book is available from the British Library.

*Special discounts are available for bulk purchases of this book. For more information,
please contact Special Sales at 410-516-6936 or specialsales@press.jhu.edu.*

The Johns Hopkins University Press uses environmentally friendly book
materials, including recycled text paper that is composed of at least 30 percent
post-consumer waste, whenever possible.

Contents

Preface

I was lucky enough to buy two great books in the 1980s. They may or may not make anybody else's list of enduring classics, and they certainly were not runaway best sellers in their time. Nevertheless, Peter Brancazio's *Sportscience: Physical Laws and Optimum Performance* and *The Bill James Baseball Abstract 1984* were inspirational. Without them, this book would not exist.

Brancazio's *Sportscience* discusses a variety of physics concepts using sports. The extent to which it teaches physics through sports as opposed to explaining sports with physics is in the mind of the reader. It does both beautifully, and it opened up an exciting new world to me. Bill James is now well known to the sports public, often referred to as the father of the statistical analysis of baseball.[1] The *Baseball Abstract* offers witty, outspoken commentary backed by a clearminded use of statistics to investigate interesting questions. In a talk at a recent mathematics conference,[2] the speaker discussed his use of the *Abstracts* to teach students how to analyze and write about statistical experiments.

Despite the large differences in subject matter, both books gave me a much-needed spark of inspiration. I have always been a sports fan but had not considered that this could benefit my career as a mathematics professor. Both Brancazio and James transferred to paper the essence of intellectual exploration—a restless desire to fully understand important aspects of sports and a pure delight in their discoveries. I read their ideas and kept thinking, "Of course, that's right. Why didn't I think of that? That's great!" The excitement of learning at the feet of two masters was invigorating.

By 1988, I had designed an interdisciplinary course on sports science, which I have taught in a variety of formats at Roanoke College. The popularity of these courses led me to incorporate sports problems into my calculus and other mathematics classes. The success of these problems was a factor in my becoming coauthor on a series of calculus books[3] and led directly to the opportunity to write this book. I hope that, in some measure, this book creates pleasure for you in the way that Brancazio and James did for me.

That you are holding this book right now means that you have overcome a powerful cultural bias against mathematics. This bias is a regrettable and, to my mind, inconceivable attitude for a country like the United States.* If you get curious about something (anything!) and start to analyze it and understand it better, you are probably doing mathematics. Mathematics follows from trying to give unbiased answers to interesting questions. The precision required to remain unbiased can sometimes be daunting. In this book, I've tried to minimize the daunt and keep the focus on the interesting questions.

An unfortunate truth of modern mathematics is that, contrary to algebra class where everything is solved exactly, simple answers are rare. This is especially true about the mathematics of golf. If you give me a golfer's swing speed, can I tell you how far the drive will go? In a word, no. It depends on what ball is being used, the properties of the shaft and clubhead, the altitude and latitude of the course, and so on. Most of what you will find here are sample calculations intended to give you a slice of golf truth,† and to improve your understanding and appreciation of the game (and, perhaps, mathematics).

---

*The word "inconceivable" is a reference to the movie *The Princess Bride,* and, yes, I do know what it means.

†My family and friends have suffered such puns for years. There is no good reason to spare you similar agony.

If you have bought this book, you have personally disproved Stephen Hawking's theory that every equation reduces the sales of a book by half.* Starting from the number of people for whom a book titled *Golf by the Numbers* might be purchased, there are enough equations in the book to reduce projected sales to one or less. My college library is required to buy a copy, so your purchase is an unexpected bonus. Thank you. More seriously, I have tried to segregate the mathematics to enhance the readability of the text. All overt calculus equations are in the endnotes.

The more technical arguments in each chapter are reserved for "The Back Tee" sections at the end of each chapter. On a golf course, playing from the back tees gives the greatest challenges and reveals the best the course has to offer. However, very enjoyable rounds can be had without venturing to the back tees. Similarly, the results in "The Back Tee" sections are interesting and accessible, but you can have an enjoyable read without facing every such challenge. My assumption is that you agree that equations are a convenient way to express precise relationships and are not a form of devil worship. I have used diagrams, graphs, and equations when I think that they help convey the information in a simpler and clearer form than the equivalent thousand words. If you so choose, you should be able to skip over the equations and find complete sentences that summarize the most important ideas to be gained from the equations.

Beyond standard high school mathematics, there is no specific mathematics course that I assume you have mastered. As in nearly every aspect of life, the more mathematics you know, the better. In my view, mathematics includes a variety of techniques for analyzing the world, so along with some equations you will see graphs, tables of numbers, logical arguments, and ideas from probability

*I do not know the origin of this rule, but Hawking uses it in the preface of *A Brief History of Time,* and I'm not averse to name dropping.

and statistics. It is all mathematics. If you are a person with a decent command of high school mathematics, a curiosity about how mathematics might be used to understand basic principles in golf, and some knowledge of golf as both player and fan, you are my ideal audience. However, even if you know little about golf, you will be able to follow along using the glossary to make sense of the arcane terms that are part of golf lingo.

As befits someone with Peter Brancazio and Bill James as muses, I have included a fair amount of physics and statistics in the book. However, I am neither a physicist nor a statistician. (My golfing buddies would be quick to jump in right now and tell you that I'm not much of a golfer, either.) Instead, I am a mathematics professor who has always done work in applied areas, incorporating physics and statistics (and biology and economics and so on) as needed. The material in the book is sometimes reproduced (with proper credit) from other sources, sometimes adapted from other sources to suit my goals (again, with proper credit), and sometimes developed specifically for this book. (In some cases, I should admit, being a mathematician makes something deeply interesting to me that might not be terribly compelling to someone else.)

I have played golf most of my life. I learned to play at Hardy's Driving Range and the numerous municipal courses in Dallas, especially Tenison Park. Unfortunately, I was born a couple of years too late to meet Hardy's most famous employee, Lee Trevino. Where I sometimes accidentally slammed worm-burning drives into the 100-yard marker, Trevino had made considerable money hitting the marker intentionally with golf balls struck with Dr Pepper bottles.[4] As well, I was too young to participate in or even know about the wild gambling that is one of Tenison Park's main legacies.[5] The high point of my golf career may have been a third-place finish in the *Dallas Times-Herald* city-wide tournament, 16-year-old division. I lost in the semifinals the day before my parents and I moved to Virginia. I played on the Virginia Commonwealth University golf team in college. (To be fair to John Rollins and the real VCU golfers, the team I played for had only club status, and all five of us who tried out made the team.) While researching material for this book, I discovered that my golf development was *mathemagenic*.[6] That is a polysyllabic way of saying that I taught myself. I benefitted greatly from the one lesson I had and envy young golfers today who have access to fantastic golf instruction and technology.

This book is not about finding "the perfect swing." Golfers my age who grew up watching Arnold Palmer, Gary Player, Lee Trevino, Doug Sanders, Miller Barber, and others have a visual library that proves there are many ways to swing a golf club well. A man named Homer Kelley[7] distilled years of intense experimentation into a complete description of the golf swing. He allowed for personal variations in many stages of the swing: more flexibility here, greater height here, extra weight there, and so on. Including variations, the perfect swing comes in 446,512,500,000,000,000 versions. (Unfortunately, the evidence shows that my swing is not one of them.)

Instead, this book is about different aspects of golf which lend themselves nicely to mathematical analysis. How much do you slice if you leave the clubface open 5°? How much extra distance do you get if you are hitting downhill? Is the handicap system biased in favor of good golfers? Is "drive for show and putt for dough" an accurate assessment of the importance of driving and putting? Who is the best putter on the PGA Tour? How accurate are pros from 200 yards out? What are the strengths and weaknesses of Phil Mickelson's game? Given that Tiger was the best on Tour in the 2000s, how much better was he than the second best? These are among the many questions that we will use mathematics to answer.

The timing is good for books like this. The computer revolution is inexorably turning the world into a more technical and quantitative place. Americans can no longer be technophobes without forfeiting a basic understanding of life. As Ian Ayres describes it in his book *Super Crunchers*[8] aspects of life ranging from wine-tasting to online shopping to public policy decisions are all increasingly governed by mathematical analysis. Ayres says, "We are in a historic moment of horse-versus-locomotive competition, where intuitive and experiential expertise is losing out time and again to number crunching." The wisdom of grizzled veterans is challenged by the results of numerical computer experiments. However, these do not have to be mutually exclusive; instead, they can enhance one another. This book in no way replaces Ben Hogan's books on golf swing principles. I view it as complementary, filling in some of the gaps that Hogan could not have known about in the pre-computing age in which he lived. Hogan, in fact, was an enthusiastic user of scientific breakthroughs for his line of Ben Hogan clubs and balls.

Golf has benefited from modern technology more than most sports. Sophisticated high-speed cameras and testing procedures have increased our knowledge of what happens when club meets ball. The lucrative industries of golf club and ball manufacturing have become highly competitive and technical. Golf telecasts have become more detailed, but too often complex statistics are thrown at us with little or no context. So, Camilo Villegas has hit 50% of the fairways and made 82% of his putts inside of 10 feet. Are those good numbers? Are those important areas in which to excel? You will find answers here which can be determined only through statistical analysis.

The use of ShotLink by the PGA has significantly raised the ante on the accuracy of golf statistics. The classic statistics of scoring average and money earned have been supplemented with greens hit in regulation, fairways hit, and others. Now we can know that in 2007 Tiger Woods averaged 27 feet, 6 inches from the hole for all shots from the fairway[9] (making him no. 1 on the Tour). Detailed knowledge of the location of each shot opens up the possibility of the development of new, meaningful statistics that can start to answer questions of course management and strategy. Perhaps you will become the Bill James of golf and fully develop the study of golf statistics.

Every year, our knowledge of the mechanics of a golf shot is stretched. New methods of analyzing a golf swing are developed. A barrage of numerical statistics is available for each tournament played. As a result, we are continuously revising our understanding of the game through the daily breakthroughs that occur in statistical analysis.

As a mathematics professor at a liberal arts college, I am sometimes asked to explain the relevance of mathematics and liberal arts in the modern world. One partial answer is this. A solid mathematical background lets you revel in the wonders of new possibilities instead of being overwhelmed by rapid technological change. A liberal arts education opens up new worlds,[10] some of which require mathematics to fully appreciate the nuances and intricacies of their residents' interactions. This book chronicles my brief excursion into the brave new world of mathematics and golf. I hope it opens up new worlds of insight for you.

I want to thank Trevor Lipscombe and the staff at the Johns Hopkins University Press. This book was Trevor's idea, and his encouragement and sound advice were crucial in helping me overcome the inevitable bumps and roadblocks that arise in getting a book to publication. Many thanks to the man "wearing the British accent." Jen Malat and Vince Burke oversaw the details of getting the text in shape to print. Kim Johnson is my fantastic copy editor. She corrected mistakes, improved explanations, reduced my use of jargon, and streamlined the structure of the text. She measured each aspect of the book against an exacting standard and helped me bring the text closer to the book I dreamed about than I could ever have achieved on my own. Thanks, Kim. My calculus coauthor Bob Smith and then-editor Liz Covello were very supportive of my desire to see this through, even though it represented a significant energy drain from our ongoing calculus work.

The general ideas that form the foundation of this book were developed in 2008, when much of the programming and analysis were done. I am indebted to Roanoke College for funding a sabbatical in the spring of 2008, which enabled me to tackle this enjoyable project.

Numerous friends were kind enough to read portions of the manuscript and provide valuable feedback. I especially thank Geoff Boyer, Brian Gray, Reid Garst, and Rich Grant for their time and helpful suggestions. Several colleagues at Roanoke College served as sounding boards for both good and bad ideas and provided much-needed expertise. I especially thank Dave Taylor, Frank Munley, and Adam Childers for friendship and advice. Thanks to Scott Berry for his generosity in sending me several of his excellent papers and for helping with my description of his work. My brother George has always been a friend and inspiration, even if he never quite made it to the green of that par 6.

The friends who have played golf with me while I worked on the book have heard more about it than anyone would ever want to. John Selby, Jeff Sandborg, and Garry Fleming are, I assume, still working on the "Sounds of Golf" CD that was to accompany this book. Thanks also to Lee Hipp, Joe Austin, Bob Stauffer, Jim Stevens, Rob Almond, and Paul Reed.

When you work at a small institution like Roanoke College, students, colleagues, and staff often become friends and supporters. Students Amanda Coughlin, Danielle Shiley, Hannah Green, and Richard Goeres worked with me on various aspects of this project. I appreciate their hard work and friendship. Linda Davis and Laura Bair both helped with numerous administrative details, and department members Chris Lee, Jeff Spielman, and Karin Saoub complete the "Trexler family" that makes work enjoyable.

Ed Parker of James Madison University and Dan Kalman of American University gave helpful feedback. I'm hoping that Ed's business venture pays off. Bruce Torrence of Randolph-Macon College helped me produce a nice article for *Math Horizons*. The late Howard Penn organized the Mathematics Awareness Month–April 2010 (mathaware.org/mam/2010/) website; the articles were later collected in a book edited by Joe Gallian.

When I started this project, I was hoping to do some statistical analysis of ShotLink data and started by typing online stats into a spreadsheet. This is not the way to go! The PGA Tour was amazingly open and helpful, and this part of the project expanded beyond my highest expectations. Thanks to Steve Evans, Senior VP of Information Systems at PGA Tour, and his executive administrative assistant, Stephanie Chvala, for their extraordinary cooperation. My PGA Tour liaison, Mike Vitti, was a pleasure to talk with and a helpful guide as I started exploring the data set.

An unexpected joy was a round at Oakhurst Links, the first golf course in the United States. Thanks to Nancy Midkiff for arranging the visit, for an excellent tour of the museum, and for helping John Selby and me get around the course in good shape. Oakhurst's owner, Lewis Keller, took time from a busy day to chat with us, giving us a sense of the history of the site and its largely unknown role in golf history. Both John and I would love to grow up to be like Mr. Keller. My "home course," Hanging Rock Golf Club, appears in a couple of places in the book. Many thanks to pros Billy McBride (who, sadly, passed away during the writing of the book) and Chip Sullivan.

Finally, and foremost, I thank my wife, Jan, and children, Kelly and Greg. After more than 30 years of marriage, Jan still encourages me to play golf. Granted, it could be to get me out of the house, but it is a true blessing to have a talented mate and best friend who genuinely wants good things to happen. As just one example, Jan researched and set up the round at Oakhurst as a surprise Christmas gift. Thank you. Kelly and Greg are not kids anymore. They are incredibly talented and good people. I'm lucky to have had a part in their upbringing and to get to follow their exploits. They will make the world a better place.

Golf by the Numbers

PART ONE

General Golf Analysis

The first part of this book introduces several ways that mathematics can be applied to golf. Since we are exploring the game in general, there are numerous conclusions that will be of use for all golfers, regardless of ability. Among the questions addressed in this first part are:

- What happens when the clubface is out of line with the target?

- How much difference does a downhill or uphill slope make?

- When putting, is it better to hit the ball hard or soft? Should you aim high or low?

- Is there any luck in putting? How much?

- Are you better off trying risky shots or playing cautiously?

- Is the USGA handicap system fair? If not, is it biased toward good players or bad players?

To answer these questions, we have developed several mathematical models. A mathematical model is a set of assumptions which (ideally) identifies in mathematical terms the most important aspects of a real world situation. The assumptions may come from physics or basic scientific principles, or they may represent a mathematical convenience. If the assumptions are realistic enough, then a mathematical analysis of the model can give us important insights into the physical situation.

ONE

The Shape of Golf

My favorite subject was always math.

—*Tiger Woods*

Tiger likes math. So does Phil. The website for the Mickelson ExxonMobil Teachers Academy quotes Phil: "As someone who uses math and science every day in my career, I recognize the importance of encouraging children's interests in math and science and equipping educators with the tools and resources they need to succeed in the classroom."[1]

This is not to say that Phil and Tiger prepare for the final round of a major tournament by solving math problems. However, it is reasonable to say that an important component of their excellence is an ongoing analysis of their swings and results. This analysis relies heavily on basic concepts from mathematics, physics, and other disciplines.

In this book, we look at various aspects of mathematics which can shed light on some of golf's most perplexing questions: Why do we miss so many putts? Why are the new drivers so big? Does the saying "drive for show and putt for dough" apply to the modern game? My hope is that these mathematical morsels will help you enjoy the game more.

We will start by examining the geometry of golf, where the mathematics is often inextricably bound to physics. A number of

excellent books on the physics of golf already exist.[2] We will borrow from them frequently as we proceed through this chapter.

On the Tee, Isaac Newton

Sir Isaac Newton (1642–1727) constructed a framework for the analysis of the motion of all things, both on and off Earth. There is no evidence that Newton himself played golf, but by the 1600s the game was already flourishing in Scotland and England.[3] Newton's second law of motion, fundamental to our understanding of the flight of a golf ball, is often expressed by the famous formula

$$\mathbf{F} = m\mathbf{a}.$$

In this amazing equation, $\mathbf{F}$ stands for the sum of all forces acting on an object (such as a golf ball), m represents the mass of that object, and $\mathbf{a}$ represents its acceleration. Acceleration, in turn, is the rate of change of velocity, and velocity gives both the speed and the direction of motion of the object. In theory, then, knowing all of the forces acting on an object lets us compute its acceleration, from which we can recover the velocity and position of the object. This idea works well enough to safely land astronauts at precise locations on the moon.

The force **F** and acceleration **a** are in boldface to indicate that they are vectors. This means that they have both size (magnitude) and direction. Newton's equation tells us that acceleration occurs in the same direction as does net force and, further, that the magnitude of the force equals mass times the magnitude of acceleration.

There are three forces that are dominant for the motion of a golf ball in flight: gravity, air drag, and the Magnus force. Gravity is the most familiar of these. In everyday language, "weight" refers to the magnitude of the force due to gravity. The weight of a golf ball is approximately 1.6 ounces and varies slightly depending on altitude and latitude. The mile-high altitude in Denver reduces weight by about 0.05% compared to sea level, while the bulging of the earth at the equator causes a weight reduction of about 0.3% moving from 50° latitude to the equator. The increased distance for drives in high-altitude places like Denver is primarily due to a decrease in air drag,[4] our second force.

Air drag is what most people think about when they hear the phrase "air resistance." It is a force that directly opposes motion, and its magnitude depends on the speed of the object. You have experienced this when you hang an arm out of the window of a moving car. You feel more air drag at higher speeds. The air drag on a golf ball depends on numerous factors, including air density[5] and the dimple patterns on the ball.

The third force, the Magnus force, affects every drive, sometimes to our embarrassment. Caused by the spinning of the ball, the Magnus force acts at right angles to both the direction of motion and the spin axis and can produce wicked hooks and slices.[6] The dominant type of spin in golf is backspin; figure 1.1 shows the directions of the three forces on a ball hit with backspin.

Imagine that the ball in figure 1.2 is moving into the page with backspin. As shown in figure 1.2a, the Magnus force is upward. If the spin is not pure backspin, but has some sidespin mixed in,

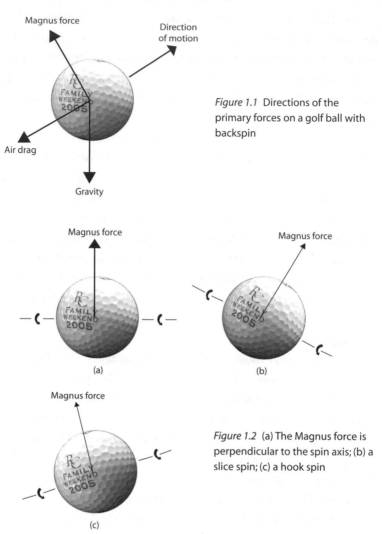

Figure 1.1 Directions of the primary forces on a golf ball with backspin

Figure 1.2 (a) The Magnus force is perpendicular to the spin axis; (b) a slice spin; (c) a hook spin

then the entire figure tilts and the Magnus force has a sideways component. In figure 1.2b, the force is up and to the right; if you are right-handed, you just hit an ugly slice. Figure 1.2c shows a Magnus force that is up and to the left, creating a hook for a right-hander.

Love Those Dimples

The usefulness of dimples on a golf ball can be explained by the drag and Magnus forces. For a golfer, drag is bad. It slows the ball down and decreases distance. On the other hand, a Magnus force is generally good. Backspin is the dominant spin on a golf shot, and the resulting Magnus force has an upward component. This causes the ball to go higher, stay in the air longer, and (usually) have time to travel farther. Figure 1.3a shows a contrast between the drag forces on a smooth golf ball and on one with dimples. Notice that the magnitude of the air drag on the dimpled ball is

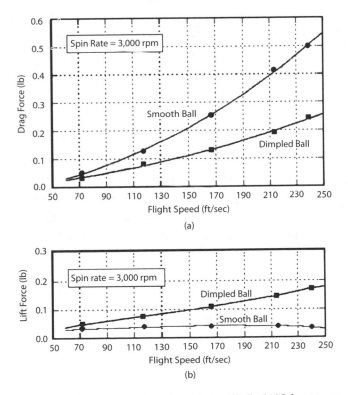

Figure 1.3 (a) Drag force on smooth and dimpled golf balls; (b) lift force on smooth and dimpled golf balls. Used with permission of Acushnet Company.

much smaller, especially at high speeds; thus, the dimpled ball travels much farther. In figure 1.3b, the upward component of the Magnus force (or lift force) is shown for smooth and dimpled balls.[7] This time, the magnitude of the force for the dimpled ball is much larger. Again, this is good. The dimpled ball goes higher, travels farther, and is easier to control.

New dimple designs are tested in wind tunnels or simulated on a computer to see if better drag and Magnus profiles are produced.

Golf Is Not a Game of Perfect Parabolas

You may have a vague memory from high school that the path of a ball in flight is a *parabola*. Like many myths recalled from our innocent youth, there is an element of truth here, but the full story is much more interesting. Parabolas are the paths of objects on which gravity is the only force.[8] This ignores air drag and the Magnus force and, therefore, is grossly inaccurate for describing most flying objects on Earth.

So why do we study parabolas in school? In a word, convenience. The "gravity-only" model produces nice equations and allows us to calculate a variety of quantities. We trade off some accuracy to keep the mathematical difficulty under control. How much accuracy do we lose? Figure 1.4 shows trajectories for two drives with launch conditions (234 ft/s is about 160 mph, typical for a professional drive) that are identical except that one ball is acted on only by gravity

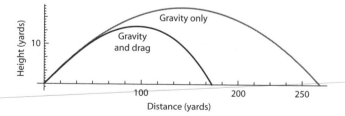

Figure 1.4 The effect of drag on a drive with initial speed 234 ft/s (160 mph) and initial angle 15°

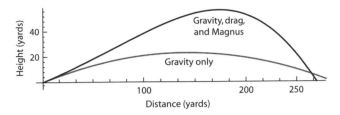

Figure 1.5 The effect of drag and Magnus force on a drive with initial speed 234 ft/s (160 mph) and initial angle 15°

and the other ball also experiences air drag.[9] The most obvious trajectory change is that the ball acted on by drag and gravity flies about three-fourths as high and less than two-thirds as far. Also, drag ruins the perfect symmetry of the parabola. The part of the curve to the right of its peak is shorter than the part to the left of the peak. This is due to the air drag continually reducing the speed of the ball. (In fact, at the end of the flight, the speed has been reduced to about 123 ft/s, just over half the initial speed.)

We can also add in a term for the Magnus force.[10] The resulting path is overlaid onto the original gravity-only graph in figure 1.5. Notice that the Magnus force propels the ball three times as high and almost as far as in the gravity-only graph, restoring most of the distance lost to air drag.[11] Carefully examine the shape of the trajectory. It starts out nearly linear, rises to a peak, and drops down, approaching a straight line at the end. This should match your experience on the golf course. Notice that the peak of the trajectory is reached about 60% of the way to the landing point. This lopsided path must be taken into account when trying to clear a tree.

As illustrated in figure 1.2 the vertical component of the Magnus force on a golf ball hit with backspin is upward. This is the source of the extra height seen in figure 1.5. An interesting consequence of this upward force is that the ball lands softly. Even without taking into account the effect of spin on a ball nestling into a lush green, this can be seen by noting that the landing speed of the ball acted upon by gravity, drag, and Magnus force is about 100 ft/s, less than half of the landing speed of the gravity-only ball.

The most dramatic consequences of the Magnus force on a golf course are hooks and slices, which are caused by sidespin. As seen in figure 1.2b, and 1.2c, a ball spinning in a plane that is not vertical will curve to the side. We'll look at some examples. To be clear, for a right-handed golfer, a "slice" is a shot that curves from left to right; a "hook" curves from right to left; a "push" has initial direction to the right of the target; and a "pull" starts to the left of target. For a left-handed golfer, all of these directions are reversed.

Drawing a Slice

The golf swing is all about delivering the clubhead to the ball in good form. Jim Furyk's loop, Bubba Watson's extension, Sergio Garcia's lag, and all of the endlessly analyzed spine angles and head bobs serve to determine the clubhead speed, direction of motion of the clubhead, and the direction in which the clubface is pointed. A mismatch between the last two creates spin. In particular, if the clubhead is not moving in the same direction as the clubface is pointing, you are likely to see a hook or a slice. Therefore, swing angle and face angle are critical. In the following discussion, the vertical motion of the ball is ignored, and we are solely interested in an overhead view of the ball's flight. The next several figures illustrate what happens if one of these angles is non-zero.

First, suppose that at impact the clubhead is moving exactly on line with the target but that the clubface is open by 5°. For a right-handed golfer, this means that the clubface points 5° to the right of the target. As shown in figure 1.6, the ball is launched to the right about 4° off-line. As seen in figure 1.7, the ball then slices well to the right and lands 260 yards down range but over 65 yards to the right of the target. The overhead view points out an interesting

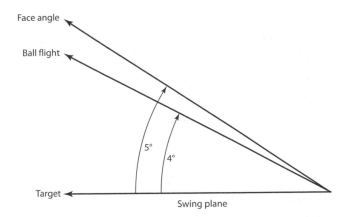

Figure 1.6 An overhead view of initial direction of ball from an open face

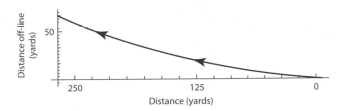

Figure 1.7 An overhead view of push/slice (ball moves right to left)

optical illusion. By the end of its flight (to the left in the figure), the ball is moving substantially off-line. In fact, at the halfway mark of the trajectory, the ball has moved only about 20 yards off-line, but by the time the ball lands, it is over 65 yards off target. In spite of this, the ball does not appear to be curving much at the end of its flight. As the ball slices, its velocity starts to line up with the Magnus force; because the Magnus force is reduced, the ball curves less. This, of course, is of little comfort to the golfer whose drive has already disappeared into the woods.

Let's look at what happens when the clubface is aligned to the target but the swing plane is not. In particular, suppose that at impact the path of the club is 5° to the left, as shown in figure 1.8.

The ball is launched slightly to the left (about 1°) of target. For a right-handed golfer, this is an outside-in swing that produces a slice. As seen in figure 1.9, the ball starts off slightly to the left and then slices back to the right, landing 264 yards down range and 44 yards to the right of the target.

If both the swing angle and the face angle are 5° to the left, the result is simply a pull (for a right-hander). Because the swing and the clubface have the same alignment, there is no slice or hook. The ball lands 278 yards down range and 24 yards to the left of the target. This drive is longer than the previous two, showing that a slice typically costs you a little distance.

The geometry in figures 1.6 through 1.9 holds in general. That is, if the face angle and swing plane are different, the ball will curve in the direction from the swing plane to the face angle. The initial direction of the ball is between the face angle and swing plane and is primarily determined by the face angle. More precisely, the initial

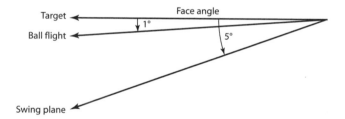

Figure 1.8 An overhead view of initial direction with swing plane to the left

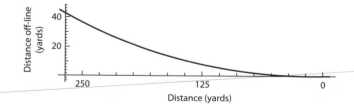

Figure 1.9 An overhead view of pull/slice (ball moves from right to left)

direction is about 80% of the way from the swing plane to the face angle.[12]

In the context of this analysis, a hook is simply a slice reversed. That is, if the swing angle is on target and the club face is pointed 5° to the left, the ball trajectory will be identical to that of figure 1.7 except it will curve to the left. The landing point will have the same down-range distance of 260 yards and will again be off-line by over 65 yards (but this time to the left). The common folklore that hooks go farther than slices is based on other factors. Hooks often have a lower trajectory that creates more roll.[13]

These calculations collectively illustrate one aspect of what teaching professionals mean when they say that the ball flight will tell you what your swing is doing. As we have seen, the initial direction of motion of the ball is primarily determined by the face angle. A pull to the left means that, at impact, your clubface is pointing to the left. Slice and hook tell you how the swing plane relates to the clubface. A slice to the right tells you that your clubface is pointed farther to the right (or, possibly, less to the left) than your swing angle. A good teacher can take this knowledge and reverse-engineer the ball flight to identify specific aspects of your swing that need attention.

A Cute Little Angle

The 18th hole at Hanging Rock Golf Club in Salem, Virginia, is a par 3 with a significant drop of over 40 feet from tee to green. My golfing buddies and I sometimes argue about how much difference the drop makes. This is our next topic.

The first surprise for experienced golfers is how small the angles in golf actually are.* For example, imagine a par 3 of 160 yards

---

*If the title of this section seems a little obtuse, you need to know that the name for a small angle is "acute."

with a 40-foot drop in elevation from tee to green. For comparison purposes, the 40-foot drop is equivalent to the height of a three- or four-story building. How steep is the slope? It will look like an impressive drop, but some triangle trigonometry in figure 1.10 shows that the angle is $\tan^{-1}(40/480)$, which is about 4.8°.

Four and five degrees may not sound like large angles, but they represent very steep slopes on a golf course.

Does 10 Yards Equal 10 Yards?

How much effect does sloping terrain have on an approach shot? Figure 1.11 shows a possible trajectory, with the ball landing on level ground at the 160-yard mark. Superimposed on this graph is a line representing a downhill slope, with a drop of 40 feet over the 160-yard horizontal distance. The intersection of the ball path with this line shows that the ball would hit the ground at nearly 170 yards, about 10 yards beyond the 160-yard mark.

As you can see, the slope creates a significant change in the distance from launch point to landing point. The downhill shot goes about a club too long. This is quantified in table 1.1.[14]

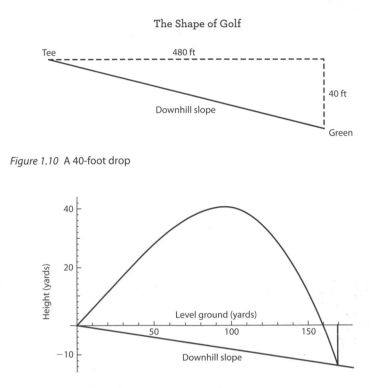

Figure 1.10 A 40-foot drop

Figure 1.11 The effect of a downhill slope on distance

If you ask golfers, you will hear different rules of thumb for judging the effects of slopes. A one-club (10 yards) difference for every 10 yards of elevation is a simple, rounded-off version. In table 1.1, a 30-foot change in elevation increases the distance by 8.4 yards going downhill and decreases distance by 9.1 yards going uphill. Rounding up to 10 yards simplifies the rule while keeping it reasonably accurate. However, there are many factors not accounted for in the calculations.

One factor that can be varied is the target distance. Using the same parameters as in figure 1.11,[15] we see that a 140-yard iron shot varies 7 to 8 yards with 10 yards of elevation change. A 180-yard iron shot varies 10 to 11 yards with a 10-yard elevation change. The longer the shot, the more the elevation shift affects the distance, even without taking into account differences in how far the ball rolls.

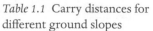

Table 1.1 Carry distances for
different ground slopes

Slope (ft)	Distance (yd)
None	160
Downhill	
10	162.8
20	165.7
30	168.4
40	170.9
Uphill	
10	157.1
20	154.1
30	150.9
40	147.8

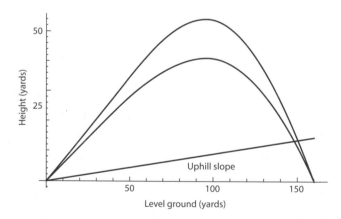

Figure 1.12 Two uphill shots: the higher trajectory hits the ground closer to the
160-yard target than does the lower trajectory

Another factor is the shape of the shot. Figure 1.12 shows two
160-yard iron shots, one the same as in figure 1.11 and the other
with a higher loft.[16] The two ball paths intersect the horizontal
axis at the same point, indicating that on level terrain they would
carry the same distance. However, on the uphill slope, the steeper

16

trajectory intersects the ground line farther to the right than does the flatter trajectory. This shows that the uphill slope reduces distance less on the high-trajectory ball than on the low-trajectory ball. The high-trajectory ball lands closer to the 160-yard target than does the low-trajectory ball. A similar result holds for downhill slopes.

Numerically, table 1.1 shows 8 to 9 yards of difference for slopes of 30 feet. For the more lofted flight in figure 1.12, the difference is 7 yards.[17] This is another popular rule of thumb: 30 feet of slope changes distance by 7 yards. The rule varies, depending on how high one hits the ball.

Another important factor for distance adjustments for irons is the orientation of the green. Werner and Greig have shown that the amount of bounce and roll after the ball lands depends critically on the tilt of the green. For flat (untilted) greens, they recommend one club for every 20 yards of elevation.[18] For greens that are tilted toward the golfer, as most are, the roll distances change and the calculations shown above are reasonably accurate.

In the past, knowing the adjustment rule was like knowing the swing weight of Tiger's clubs. While it might be interesting, it would not help your game. Standing on the tee, unless you could judge whether a drop in elevation was 20 feet or 40 feet, all of the rules were useless. Nowadays, however, some range-finders display adjustments for slope. If you use such a device, pay close attention to its accuracy and determine if the range-finder's rule is based on the shape of somebody else's shots. You may need to adjust the adjustment.

The Back Tee: Equilibrium

We close the chapter with a detailed look at the angle of flight of the ball. In particular, at what angle does the ball hit the ground? While this can vary from shot to shot, we can gain important information from an equilibrium calculation.

Equilibrium is a basic property of any system that mathematicians and engineers explore. An equilibrium is a balance point where the variables that are being tracked do not change. Equilibrium values may also serve the role of being the natural values that occur in the absence of external forces. For example, an equilibrium value for a bowl is the bottom of the bowl. If a ball starts somewhere in the bowl, a reasonable prediction for where the ball will be a minute later is at the bottom of the bowl. Unless someone flicks the ball or the bowl is turned upside down, the natural position of the ball is at the bottom. Some greens have "collection points" that operate this way. A ball that starts near the collection point will end up at the collection point.[19] What about the equilibrium of the angle of flight?

Returning to the downhill par 3 shown in the photo of the 18th hole at Hanging Rock, on several occasions I have been in a group that disagreed about how much difference the drop makes. One golfer says two clubs, another thinks between one and two clubs, and I usually claim one club. Consistent with the calculations shown earlier in this chapter, the three actors in this drama have different ball flights. The two-club advocate hits a flat ball, I have the highest trajectory, and the one- to two-club advocate is in between. So we could all be correct.

Mathematicians look for patterns. Inventing and solving little mental puzzles is the modus operandi of the mathematical mind. An interesting thought experiment here is to imagine a golfer who hits irons much higher than I do. Based on the pattern discussed above, this golfer would need to adjust less than a full club on our downhill par 3. Now, extrapolate this to a mythical golfer who hits the ball almost straight up. Does the mythical golfer have to make *any* adjustment? Stated differently, is there a trajectory so steep that when you superimpose it on figure 1.11 there is essentially no difference in horizontal distances?

The answer is no. The reason is backspin. The right angle formed by the direction of motion and the Magnus force is maintained throughout the flight of the ball. If the ball is on the way up, as in figure 1.1, the Magnus force from backspin points up and back. If the ball is coming down, the Magnus force is up and forward. Imagine rotating the direction of motion in figure 1.1 by 60° clockwise. The Magnus force and air drag directions also rotate by the same angle, so that the Magnus force has a forward component (to the right in the figure). If the ball is moving straight down with backspin, the Magnus force is forward. When the Magnus force pushes it forward, the ball is no longer moving straight down. So, if there is backspin, the ball cannot literally drop straight down, but can it maintain a near-vertical drop? This is an equilibrium question.

If there is an angle that the trajectory maintains on the way down, then that angle corresponds to an equilibrium value for the slope of the trajectory. In figure 1.11, it does look like the graph has straightened into a line with constant slope.

An equilibrium value for the slope of the ball trajectory can be found by solving an equation obtained by setting the sum of the forces on the ball equal to zero.[20] Actually, the equation is ugly enough that we can only estimate the solution. For a constant spin rate of 5,000 rpm, the equilibrium slope is approximately −0.92. This makes an angle of about 43° with the horizontal, which is not even close to being vertical! By comparison, a constant spin rate of 3,000 rpm has an equilibrium at 47° from the horizontal. This is steeper than 43° but still far from the 90° vertical mark.

Do golf balls actually reach these equilibrium values? Thinking that 43° looked right for figure 1.11, I created figure 1.13 confident that the slope would slide down to −0.92 (43°) and level off. This figure shows the slopes of a trajectory for an iron shot hit from an elevation of 50 feet with a spin rate of 5,000 rpm.

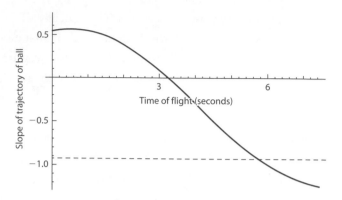

Figure 1.13 The slopes of the trajectory of an iron hit from an elevation of 50 feet, passing through the equilibrium value of −0.92

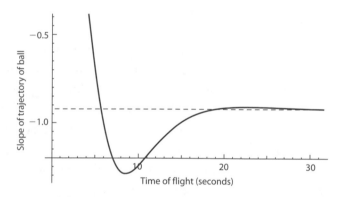

Figure 1.14 The slopes of the trajectory of an iron hit from an elevation of 2,000 feet, eventually approaching the equilibrium value of −0.92

The lower horizontal line shows the equilibrium value of −0.92. Instead of the slopes leveling off at the equilibrium, they just blow right through what I expected to be a barrier! Fortunately, the computer allows experiments corresponding to ridiculous situations. Figure 1.14 shows the slopes for a trajectory of a shot from an elevation of 2,000 feet.

Given long enough to fall, the slopes do approach the equilibrium value. The result applies only to a hole you would see on an "Impossible Golf Holes" calendar, but the math works.[21]

As a final comment on this investigation, I should confess that the assumption that the spin rate remains constant is not exactly valid. Although the actual rate at which the spin decreases is not agreed upon (see note 10), if the spin rate changes by any reasonable amount, there is no single equilibrium value to be approached.

The theoretical lesson to take from this is that for a typical iron shot, whether you hit the ball high or not, the ball drops onto the green at an angle reasonably close to 45°. The practical lesson is somewhat different. Small differences in angles due to swing mechanics or equipment make a large difference in the distance carried and in the amount of roll. The higher the trajectory, the less difference the slope of the ground makes. The adjustment that you should make depends on your ball flight and may be different from the adjustments of your playing partners.

TWO

Golfer's Spread
Variation in Golf

The only thing constant about golf is its inconstancy.

—Jack Nicklaus

Many golfers, including Sam Snead and Lee Trevino, considered Moe Norman to be the best striker of the ball in golf history.[1] Tiger Woods has said that only two players have ever truly "owned their swing"—Moe Norman and Ben Hogan.[2] Yet, many people have never heard of Norman, whose story is fascinating. Raised in Ontario in the 1930s, he latched onto golf as an escape from social and family difficulties. He taught himself a unique, highly repeatable swing that utilized heavy clubs, a very wide stance, a grip anchored in the palm of his right hand, and a single plane for backswing and downswing. It is closely related to the Natural Golf method.[3]

In spite of amassing staggering statistics—like 33 course records, 17 holes-in-one, and three rounds of 59—Moe Norman never made it on the PGA Tour. He had a number of, to put it nicely, eccentricities. Opponents carefully reaching out to mark their balls on the greens might find Moe's putt scurrying underneath their arms. Moe liked to play fast. A playing partner bemoaning his drive of a brand new ball into a creek was startled to find Moe knee-deep in the water trying to retrieve the ball. Moe had finely tuned ball-hawking skills from his days as a caddie, appreciated the value of a new ball, and was a very loyal friend; however, he dressed poorly

and was ill-at-ease socially. While some reporters loved having a character on tour, others thought he was not enough of a gentleman. He was made to feel unwelcome on the Tour and returned to Canada.[4]

If we say that Moe Norman was the best ball striker, what exactly do we mean? To a mathematician, the word "best" is not meaningful without a precise definition. Are there objective criteria by which we can say that Moe was better than Ben Hogan* or Jack Nicklaus? While distance and height, and the ability to hit controlled draws and fades, are part of ball striking, the primary criterion for ball striking is consistency.

---

*In *Talking on Tour*, Don Wade tells about Ben Hogan filming a television commercial in which he hit a ball from the fairway. When the director instructed Hogan to hit it onto the green, Hogan replied, "What would you like it to do when it gets there? Bounce left, right, or back up?"

Measuring Consistency

Don Wade tells the story of Moe Norman arriving at the site of a tournament and finding the course blanketed in fog. Instead of feeling his way to the driving range, Moe grabbed a handful of balls and drove them down the first fairway. The caddie who went to retrieve the balls was astonished to find two balls touching each other and all six balls within a 10-foot circle.[5]

This story gives us a place to start. Imagine two golfers hitting six balls each at the driving range. To keep things simple, let's assume that all of the balls end up in a line, as illustrated in figure 2.1. Obviously, golfer A is more consistent than golfer B, since B's shots are more spread out, but how much more consistent is golfer A? A basic measure used by statisticians to quantify spread is *standard deviation*.

The formula for the standard deviation σ of a data set is often written in the form

$$\sigma = \sqrt{\sum_{i=1}^{n} \frac{(x_i - \bar{x})^2}{n - 1}},$$

which is a little like legal fine print: you know that it's good, but it doesn't seem to be in a recognizable language. Taken in the right

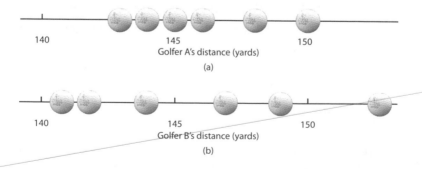

Figure 2.1 Measuring consistency: (a) golfer A, more consistent; (b) golfer B, less consistent

order, the standard deviation calculation is quite simple. First, find the mean (average) of the data; this is denoted $\bar{x}$. Golfer A has distances 143, 144, 145, 146, 148, and 150, and the mean is

$$\bar{x} = \frac{143 + 144 + 145 + 146 + 148 + 150}{6} = 146.$$

Golfer B has distances 141, 142, 144, 147, 149, and 153, and the mean is

$$\bar{x} = \frac{141 + 142 + 144 + 147 + 149 + 153}{6} = 146.$$

The next step is to subtract the mean (146) from each data point.

A: 143, 144, 145, 146, 148, 150 $\rightarrow$ $-3, -2, -1, 0, 2, 4$

B: 141, 142, 144, 147, 149, 153 $\rightarrow$ $-5, -4, -2, 1, 3, 7$

Square these numbers and add them up.

A: $9 + 4 + 1 + 0 + 4 + 16 = 34$

B: $25 + 16 + 4 + 1 + 9 + 49 = 104$

Divide by $n - 1$, which is one less than the number of data points (in this case, divide by $6 - 1 = 5$), and then take the square root.

$$A: \sqrt{\frac{34}{5}} \approx 2.6 \quad \text{and} \quad B: \sqrt{\frac{104}{5}} \approx 4.6$$

The standard deviations are approximately 2.6 for golfer A and 4.6 for golfer B. This gives us a measure of how much more consistent golfer A is than golfer B, as the spread in golfer B's shots is nearly twice the spread in golfer A's shots.

Notice that, for each golfer, four of the six data points are within one standard deviation of the mean. That is, for golfer A there are four balls in the distance range 146 ± 2.6, between $146 - 2.6 = 143.4$

and $146 + 2.6 = 148.6$. For golfer B, four of the six distances are in the range 146 ± 4.6. This illustrates a guideline that can be made precise: for samples from a normal (bell curve) distribution, about 68% (roughly two-thirds) of the data should be within one standard deviation of the mean; about 95% of the data should be within two standard deviations of the mean; and 99.7% should be within three standard deviations of the mean.

Standard deviation is a quantity that will resurface in this book several times. In golf and numerous other situations, it is a simple and informative measure of the amount of variability in a set of data.

The Golfer's Spread

In figure 2.1, we ignored the left/right movement of the ball. We will now factor in this second dimension of motion. The question is what the result would be if a golfer dropped several balls and hit them all with the same club at the same target. If the golfer is Moe Norman, you might imagine a neat pyramid of balls stacked on top of the target. By contrast, a large park would be required to hold my shots. What would the scatter of balls look like?

All mathematical investigations begin with assumptions. In what follows, I assume that every swing is the same except for random variations in swing speed, swing angle, and face angle. These variations will be assumed to be independent—that is, the size of the variation in swing speed has no relationship to the variation in swing angle or face angle. They are not really independent—a particular twitch in the downswing could simultaneously pull the swing plane to the left *and* open up the club face—but the calculations are easier with this assumption. The remaining question is how much variation to allow. The results of $5°$ misalignments are shown in figures 1.7 and 1.9 of the previous chapter. Is $5°$ too large or too small?

Researchers Werner and Greig estimate a standard deviation in swing angle for scratch golfers of $2.314°$.[6] Thus, an error as large as

5° is greater than a two-standard-deviation error and would occur less often than 5% of the time. For non–scratch golfers, the standard deviation increases by 0.0673 times the handicap. For example, a 20-handicapper would have a standard deviation estimated at $2.314 + 20 * 0.0673 = 3.66°$. For this golfer, a 5° error is between one and two standard deviations and would be likely to occur every few holes.[7]

The standard deviation in clubface angle for scratch golfers is estimated at 1.42°, with an adjustment of 0.0412 times handicap for other golfers. For example, a 10-handicapper would have a standard deviation of approximately $1.42 + 0.412 = 1.832°$ for clubface angle. The standard deviation in swing speed is estimated as 2.984% plus 0.0868 times handicap.

The computer experiment simulates shots by a scratch golfer. The starting point is a basic swing with clubhead speed equal to 161 ft/s (about 110 mph, the Tour average) and both angles set to 0. The landing point is found and then an estimate is made of the stopping point after roll.[8] Two more speeds are generated: $161 + 9.608$ ft/s and $161 − 9.608$ ft/s. The change of 9.608 corresponds to two standard deviations, where one standard deviation is 2.984% of 161, or $161 * .02984 = 4.804$ ft/s. Similarly, swing angles of ±4.628° and face angles of ±2.84° are generated. There are 27 combinations of one of the three swing speeds, one of the swing angles, and one of the face angles. Stopping points for each of the 27 sets of initial conditions are estimated, and they are shown in figure 2.2.

The dots in figure 2.2 represent the computed stopping points. A solid polygon has been drawn around the points to help visualize the region where all good shots would be. In this case, a "good" shot means that each of the three variables is within two standard deviations of its base value. Assuming that the variables are independent, about 86% of the shots would be inside of this polygon. (This is obtained by using the two-standard-deviation rule on each of the

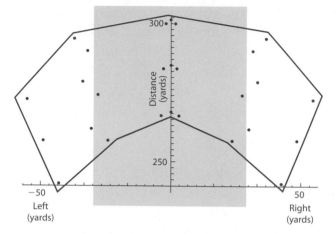

Figure 2.2 Stopping points for shots from 27 swings of different speed, face angle, and swing plane

three variables: $0.95^3 \approx 0.86$.) The shaded region represents a 60-yard-wide fairway and is included for reference.

Loosely speaking, the points in figure 2.2 form three arcs corresponding to the different swing speeds. A 20-handicap golfer would show a similar pattern but with more scatter and, therefore, fewer balls in the fairway.

Figures 2.3a and 2.3b show the corresponding scatters with the basic shot being a fade[9] and a draw, respectively. There are fewer balls in the fairway in these cases compared to the "straight" drives in figure 2.2. This result depends strongly on the orientation of the fairway. A straight fairway calls for a straight drive.

Suppose that the fairway curves to the right or left. You can fit more of the dots in figure 2.3b (from a draw) into a fairway that curves from right to left than a fairway that curves left to right. This illustrates the standard wisdom of shaping the shot to the shape of the fairway. However, the difference in the number of balls in the fairway is not large. For hitting fairways, the most important factor by far is the size of the spread, which is determined by the consistency of the ball striker.

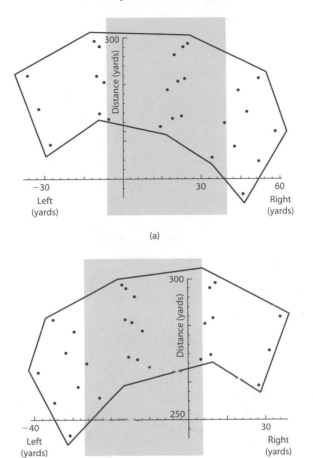

Figure 2.3 Stopping points for 27 shots with (a) a basic fade swing and (b) a basic draw swing

The shot pattern for irons will have the same general shape as that seen in figures 2.2 and 2.3. Instead of staying within a strip of fairway, of course, the goal is to be on the green. Clearly, the shape of the green will have a large effect on how many balls stay on the green.

We next examine what happens once you're on the green and ready to putt.

A Polar-ized Lens

When golfers think about putts, there are usually two properties that come to mind: speed and line.[10] That is, we have control over how hard we hit a putt and in which direction we hit it. If we start the ball on its way with the correct initial speed and direction, the putt will go in . . . unless the golfing gods are angry with us (the topic of the next chapter). Mathematically, we describe putts in terms of polar coordinates.

The mathematical position of an object in two dimensions is usually described in relation to a fixed location called the origin. In rectangular (two-dimensional) coordinates, we determine the object's horizontal distance x and vertical distance y from the origin and call the location (x, y). In polar coordinates, the object is defined by its distance r from the origin and the angle θ from the positive x-axis, as shown in figure 2.4. Mathematically, the x and y version of a point is so familiar that polar coordinates can seem odd. For putting, though, polar coordinates are natural, since they identify distance and direction.

For this reason, Werner and Greig use polar coordinates to describe some important results from their research.[11] To gather data

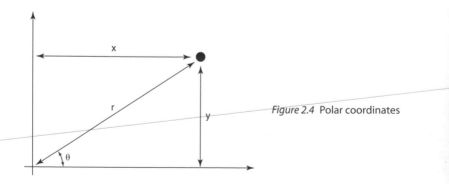

Figure 2.4 Polar coordinates

on putting accuracy, they recruited numerous golfers of varying handicaps. Each golfer was given a small target (not a hole) at which to aim, on a level green from varying distances. The actual stopping points of the ball were recorded for a large number of putts. Averaging over numerous golfers, Werner and Greig report that the standard deviation for distance is

$$\sigma_r = (0.0536 + 0.0017 * \text{HCP}) * D,$$

where D is the target distance and HCP is the player's handicap.

To put some numbers on this, start with a 15-foot putt by a 10-handicapper. Then $D = 15$ feet, HCP $= 10$, and you can compute $\sigma_r = (0.0536 + .017) * 15$ ft $= 1.059$ ft. This says that about two-thirds of the putts will roll a distance in the interval 15 ± 1.059 feet and, therefore, end up in the range from 14 feet to 16 feet. About 95% of the putts will be within two standard deviations of the mean, roughly between 13 feet and 17 feet. Remember that this calculation applies to *flat* putts that the golfer is trying to hit exactly 15 feet. The final 5% of putts are either more than 2 feet short of or more than 2 feet beyond the target. This is part of what separates a 10-handicapper from a scratch golfer.

The standard deviation for angles is estimated as

$$\sigma_\theta = (1.718 + 0.0039 * \text{HCP}) \text{ deg}.$$

Applying this formula to a 10-handicapper gives $\sigma_\theta = 1.757°$. About one-third of the putts will have an angle that is at least $1.757°$ off. Over 15 feet, this translates to at least 5.5 inches off-line, more than enough to miss the putt.

Combining this information gives a picture of where the putts finish. Figure 2.5 shows the stopping region for a putt of 15 feet by a 10-handicapper. A circle is drawn at the 15-foot mark to add perspective on the size of the hole. The shaded area covers ± 1 standard deviation for distance and ± 1 standard deviation for direction. Assuming that the errors are independent (unfortunately they are

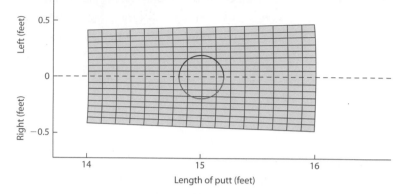

Figure 2.5 Stopping points for 15-foot putts by a 10-handicapper, assuming that distance and angle are within 1 standard deviation of the target. The ball moves left to right, and the hole is shown as a circle.

not), that means only about 46% ($0.4624 = 0.68^2$) of the putts will be inside the shaded area, and over half of the putts will be outside of the shaded region.

This raises an interesting question: which of the stopping points in figure 2.5 corresponds to putts that would go in? Certainly every stopping point inside the hole would count. However, what about a point just to the right of the circle? The figure corresponds to a putt that travels from left to right. A putt that is at the center of the hole but hit an inch too far will go in. So, many of the points that are outside of the circle represent putts that would go in. The "make zone" examined next provides more details.

Entering the Make Zone

The most important question about a putt is whether the ball goes in the hole or not. As all golfers know, having the path of the putt intersect a portion of the hole does not guarantee that the ball goes in. If the ball has too much speed, it will roll right over the hole, or "lip out." On the other hand, the putt does not have to be perfect. Some putts pound the back of the hole and drop in, and there is the "all-around good" putt that does a lap or two around the hole before dropping. The make zone is a way of illustrating which putts go in and which do not.

One make zone is based on where the putt would stop if there was no hole. Werner and Greig have developed formulas for the make zone for different speed greens, based on extensive data collection.[12] Hoadley found a similar shape.[13] In both cases, the technique is to roll balls on flat putts from a fixed distance with a variety of initial speeds and lines. For the putts that go in, the same speeds and lines are repeated on a flat putt with no hole present, and the stopping point is determined. The make zone is an idealized region showing where made putts would have stopped if the hole had not been in the way.

Figure 2.6 shows the make zone for a flat green with the putt moving from left to right and the hole centered at the crossing of the axes. If you hit a putt at the center of the hole, it will drop even if it has enough speed to roll 52 inches (a little over 4 feet) past the hole. However, a putt that is just inside the edge could have only enough speed to roll 10 inches or fewer past the hole.

The portion of the make zone to the left of the vertical axis in figure 2.6 is a semicircle, representing the front half of the hole. Its radius is larger than the radius of the hole. To see why, you need to know that in this analysis the "location" of the ball is actually the location of the center of the ball. So, what should happen if the ball stops just outside the edge of the hole? The question refers to the center of the ball being just off the edge, which means that 45% or so of the ball is hanging over the hole. I would certainly want

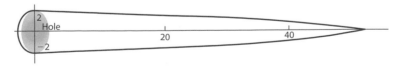

Figure 2.6 Make zone for a flat putt rolling from left to right

that putt to drop. And, in fact, it usually does. Werner and Greig estimate the radius of the make zone semicircle to be $2.125 + 1.944/S$ inches, where S is the Stimp number[14] of the green in feet. The actual radius of the hole is 2.125 inches, so the expression $1.944/S$ gives the leeway for a ball dropping in. For a fast green with Stimp 10, you get only an extra 0.2 inches. For a slower, hairier green of Stimp 5, the leeway is close to 0.4 inches, which is about half the radius of the ball. For a putt dying at the hole, this means that slower greens are more forgiving.

A different way of constructing a make zone is to show the combinations of initial speed and direction that produce a made putt—that is, the variables that the golfer directly controls. However, this version of the make zone is highly dependent on the length of the putt, whereas figure 2.6 is valid for any length of putt. The make zone in figure 2.7 is for a flat 15-foot putt.

The shaded region in figure 2.7 indicates the values of the initial speed (in ft/s) and initial angle (where 0 is directly at the hole) which will produce a made 15-footer, based on the make zone in figure 2.6. Notice that there is less than a 1° margin of error on either side of center. Reading off of the vertical axis, you can see that even for

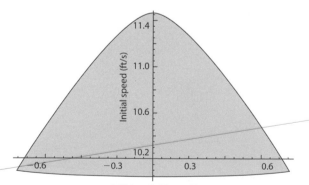

Figure 2.7 Combinations of speed and initial direction to make a flat 15-foot putt

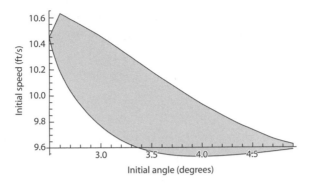

Figure 2.8 Combinations of speed and initial direction to make a downhill, right-to-left, 15-foot putt

a putt hit in the center of the hole, the margin of error for speed is less than 1.5 ft/s. It takes a precise stroke to make a 15-footer!

The symmetry in figure 2.7 depends on the putt having no break. Figure 2.8 shows the analogous make zone for a putt with break. The ball is still 15 feet from the hole, and the speed of the green has not been modified. The green is assumed to be a flat plane tilted at an angle of 2°, such that the putt is downhill and breaks right-to-left.

To read this graph, you can start with an angle. For example, 3.5° means that the putt is aimed 3.5° to the right. This converts to allowing for 11 inches of break.[15] Finding 3.5 on the horizontal axis, you see that the shaded region extends from just below 9.6 (the actual value is 9.57) to about 10.2 (more precisely, 10.19). The putt will be made if the initial speed is between 9.57 ft/s and 10.19 ft/s. An initial speed of 9.57 ft/s barely gets the ball to the hole, whereas an initial speed of 10.19 sends the ball to the back of the hole, almost producing a lip-out. The graph can give an answer to the question of what the best line is. There are numerous ways to answer this question, depending on how you define "best." One goal might be to have the maximum margin of error in speed. At angle 3.2°, speeds between 9.67 ft/s and 10.35 ft/s are all good. The difference of 0.68 ft/s is the margin of error, which is the largest such margin of error possible. By contrast, notice that the angle of 3.5° only produces a margin of error of $10.19 - 9.57 = 0.62$ ft/s.

Unfortunately, it is not helpful to tell a golfer to use an angle of 3.2° and any initial speed between 9.67 ft/s and 10.35 ft/s. Protractors are not standard golf equipment, even in the well-accessorized golf bag. However, this result can be translated into golfing terms. At the angle of 3.2°, all made putts enter the hole with a fair amount of speed. The extreme speeds of 9.67 ft/s and 10.35 ft/s are right at the values for lip-outs on the low and high sides of the hole, respectively. The conclusion is this: if you want the largest range of initial speeds possible, use a relatively small angle and *hit the putt firmly*. The downside is that the high speed will take the putt well past the hole if you miss. If you do not want the agony of a long second putt, then you should choose a larger angle and slower speed. Figure 2.8 shows that this safe play gives a slightly smaller margin of error. We will revisit this trade-off between aggression and safety in the next chapter.

Bad Breaks

Should you aim higher or lower on a breaking putt? I have never felt terribly comforted by the comment, "At least you missed on the high side. That's where pros miss." However, there are a couple of factors that point to "aim higher" being good advice. First, most golfers underestimate the break.[16] (This is discussed further below.) Plus, if you lag the putt to finish near the hole, you will need to play much more break. In figure 2.8, the angles range from 2.5° to 4.9°. These angles correspond to breaks of about 8 inches and over 15 inches, respectively. The amount of break you play can change drastically with the firmness of the putt.

The ability to read the amount of break on a putt involves an interesting mental illusion. Golfers talk about having a "feel" for the green, and this vague term is appropriate. It turns out to be difficult to quantify the ability to read greens. For example, if you say that a putt will break 5 inches, what exactly does that mean?

Suppose that the bottom of the box in figure 2.9 represents a distance of 5 inches below the center of the hole. Eyeballing this graph, would you say that this putt breaks more or less than 5 inches? What most golfers mean when they say a putt breaks 5 inches is that the ball should initially move directly at a point even with the hole and 5 inches to the side. By this measure, the putt in figure 2.9 breaks nearly 8 inches. Surprised? The graph and calculation in the notes may help.[17]

Short-game expert Dave Pelz has found that most people severely underestimate the break of a putt. However, this may be partially due to the optical illusion demonstrated in figure 2.9. Translating Pelz's data to a putt that breaks 20 inches, on the average golfers report that they think the putt will break a mere 5 inches but line their putter head up for a 15-inch break and then "mishit" the putt on the line for an 18-inch break.[18] Some of these putts go in, so in this sense the read is not far off. Much of the error is in the mental measurement of where the ball is actually being aimed. The unusual, and apparently intentional, mishit of the putt may be part of why sharp-breaking putts are so difficult for amateurs to make.

For the average golfer, then, reading a green is more of an intuitive visualization than a precise calculation. This is why getting the feel of a green and visualizing the path of a putt going in are so important.

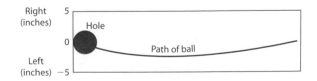

Figure 2.9 Overhead view of a sidehill putt: how much did it break?

Sources of Spread

The graphs in this chapter have shown some of the effects of variation in golf. But where does the spread come from, and how can you control it?

For drives and iron shots, most of the variation comes from our swings. We have seen how misaligned swing planes and face angles produce pulls, hooks, pushes, and slices. Also, variations in swing speed affect ball speed and spin. Similarly, there are three main sources of error in a putting stroke. The swing plane can be misaligned, the face angle can be misaligned, and the ball can be hit off-center. Dave Pelz has carefully researched the effects of these errors.[19]

Only about 20% of the swing plane error is transmitted to the ball. That is, if the swing plane aligns 10 inches to the left of target, the ball will go 2 inches left of the target. This is good news for those of us with outside-in putting strokes and explains why Billy Mayfair could win tournaments with a stroke that followed no known path.

On the other hand, a full 90% of face alignment errors are transmitted to the ball. If the putter face points 10 inches left of the target, the ball will go 9 inches left of target. If you have ever helped someone line up a putt, you probably focused on face alignment, which is the single most important factor in lining up correctly.

If you think that hitting putts in the center of the club should not be difficult, you are not alone. However, Pelz finds that this is a common and devastating error to make. When a putt is hit off center, the putter head twists. The twisting affects the initial direction and speed of the putt so that the actual putt bears little resemblance to the intended putt.

La Revolution, C'est MOI

The primary sources of our golfer's spread are our own swing errors. Our swings can be improved with practice and good instruction. Along with practicing more, we can reduce much of the spread with better clubs. The principal ingredient in this technological diet pill is MOI, or *moment of inertia*. Without getting into technicalities, the moment of inertia for a particular object and a particular rotation axis is a measure of the object's resistance to rotation about that axis. The larger the MOI, the less likely the object is to rotate.

Golf club manufacturers can now make clubs that, while light, have large MOI's for common rotations. The gigantic 460-cc driver clubheads, perimeter weighting on irons, and branding-iron shapes of putters are all intended to increase MOI. With less twisting of the club on off-center hits, we get more consistency on our shots.

The Back Tee: Streaking Tiger

The graphs in this chapter illustrate the importance of consistency for all golfers. The more you can reduce the spread of your shots, the better you will score. How important is consistency at the top level of the game? The following study, conducted by Richard Goeres,[20] examines one aspect of consistency in the professional game. In this discussion, we will be looking at consistency in scoring for a round, as opposed to consistency on a particular shot.

From 1998 to 2005, Tiger Woods made the cut in every tournament he played. His streak of 142 consecutive cuts made is the official PGA record.[21] Goeres wanted to get a handle on how impressive this record is. In the study, a field of 110 golfers was generated randomly by a computer. Each golfer was assigned a "talent level" chosen from a normal distribution with mean 70.8 (the Tour average for the 1998 through 2005 seasons) and standard deviation 0.7. Scores for two rounds for each player were then simulated, from a normal distribution with mean equal to the player's talent level and standard deviation of 2.5 (the standard deviation in scores for a typical round in a PGA tournament). The cut was set at the top 70 plus ties.

Tiger's mean was first set to 68.5 (his scoring average from 1999 through 2003), meaning that he would be more than 2 strokes per round better than average and a full stroke better than 97% of the golfers he might face. With the same standard deviation of 2.5 strokes for a round as the rest of the field, the simulated Tiger missed the cut 3.84% of the time.

With this percentage, the likelihood of making 142 straight cuts computes to about 1 in 260. To be precise, the probability of Tiger making a cut is assumed to be 0.9616. Assuming that each tournament is an independent event, the probability of making 142 out of 142 cuts is 0.9616^{142}, or approximately 0.004. The fraction $\frac{1}{260}$ evaluates to approximately 0.004.

To get a handle on this, some graphs may help. In figure 2.10a, the curve illustrates the distribution of talent levels for the simulation.[22] Tiger's talent level of 68.5 is marked for reference. Figure 2.10b shows the distribution of Tiger's simulated scores, along with the distribution of simulated scores for an average golfer in the field. Since making the cut in this study only required being in the top 70 of 110 (many fields are larger), the advantage given to the simulated Tiger would seem to guarantee a made cut. However, it turns out that there is enough overlap in the distributions to produce missed cuts nearly 4% of the time.

In a different simulation, Tiger's mean for the week was set equal to the best of the randomly generated means for the field of 110. His standard deviation was lowered by a full stroke to 1.5. Examining figure 2.10a, we can see that, if one of Tiger's competitors drew a talent level in the left tail of the distribution (in particular, more than 3 standard deviations to the left), Tiger's talent level could be the same 68.5 as in the first simulation. In many cases, Tiger's talent level was higher in these tournaments than in the first simulation. Being rated as the best player in the field and being far more consistent than anybody in the field, Tiger would see his likelihood of making 142 straight cuts

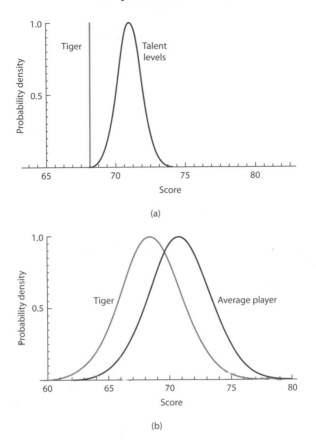

Figure 2.10 Consistency in scoring for a sound: (a) talent levels (average scores) of simulated field; (b) score distributions for simulation

increased to about 1 in 7—still not very likely, but not all that shocking.

Notice that when Tiger's mean was raised (slightly, on the average) and his standard deviation was lowered by a stroke per round, his odds of making 142 straight cuts went up dramatically. The consecutive cut streak is not likely if Tiger is just better than everybody else. Along with being the best player in the field every week, he must play closer to his average than everybody else to keep making cuts. The graphs illustrate this result. Figure 2.11 shows the

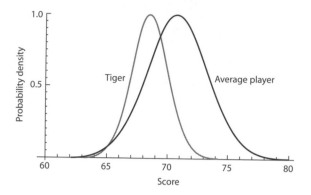

Figure 2.11 Score distributions for second simulation

distribution of Tiger's simulated scores with an average of 68.5 and a lower standard deviation, along with the distribution of simulated scores for an average golfer in the field. Comparing figures 2.11 and 2.10b, you may not immediately conclude that the streak is more likely in the situation of figure 2.11. The key is to realize that only a bad round by Tiger would put the streak in jeopardy. In figure 2.10b, Tiger's bad rounds extend out to 75 and beyond. In figure 2.11, a score higher than 72 is very unlikely. Therefore, with the lower standard deviation there was little chance that Tiger would have two rounds bad enough to put him in danger of missing the cut.[23]

The simulations point to Tiger's consecutive cut streak being very special. There are other ways of reaching the same conclusion. As of October 2010, the longest active streak of made cuts was Steve Stricker's 23 straight cuts. He was the only player who had made cuts in more than 20 consecutive tournaments and one of only three who had made 15 consecutive cuts. The record of 142 looks very, very good.

THREE

Good Luck Putting
Randomness on the Greens

I hit it perfectly, but it wouldn't go in.

—*Tom Kite*

The image is iconic. Jack Nicklaus's 12-foot putt to take the lead on the 71st hole of the 1986 Masters is one of the most replayed shots in golf history. Along with its critical role in Nicklaus's record sixth win, this putt has been canonized because of a perfect camera angle and an inspired three-word call by announcer Verne Lundquist. Lundquist gave us a hopeful "maybe" as the ball crawled toward the hole and then, when it dropped, a triumphant "Yes, sir!" beautifully synchronized to Nicklaus's celebratory two-arm vertical thrust.

Lundquist's initial "maybe" shows that he knew how tricky this putt was. Nicklaus's caddie and son, Jackie, initially read the putt to break to the right, but Jack thought that this would be offset by the tendency of putts at Augusta National to break toward Rae's Creek, which was to the left. The putt did indeed start to move right before straightening out. However, Jack has said, "I've gone back and putted that putt again a hundred times . . . I've never found that ball to go left again. But it did that time."[1]

Not everyone was as fortunate as Nicklaus that day. Playing in a later group, Tom Kite experienced the agony of defeat. His 18th hole putt to tie Nicklaus and force a playoff refused to break enough and slid agonizingly over the edge of the hole. Whether these events were luck or fate is not entirely a philosophical toss-up.[2] The role

of luck in putting is the topic of this chapter. As we will see, there is a surprisingly large element of chance in making a putt, even on the nicest greens in the world.

Bad Putting?

Here is a statistic that surprises almost everyone, including professional golfers. What percentage of 12-foot putts does a professional golfer make? If you haven't already heard the answer, think about it for a minute and take your best guess. Data collected at PGA tournaments in 1988 showed that the pros made a mere 25.7% of their 12-foot putts, with data collected only on *flat* greens.[3] This is an amazing statistic. The best golfers in the world missed almost three-fourths of their 12-foot putts! By 2009, the percentage of putts made from 12 feet had increased to about 32% on all greens. Percentages from other distances are given in chapter 7, as well as breakdowns of which pros are the most succesful from different distances.

Putting guru Dave Pelz found one cause for the pros' struggles. He built a machine called the True Roller that reliably reproduced both line and speed on a putt.[4] The True Roller is basically a ramp, carefully engineered so that the ball rolls smoothly down the ramp and onto the green without bouncing. Pelz can aim the ramp in whichever direction he wants, and the speed of the ball on the green is determined by how far up the ramp the ball starts. The True Roller was tested to verify that its putts did not vary in direction or speed.

Pelz set up the True Roller at Westchester Country Club on pro-am day of the PGA's Westchester Classic tournament on a freshly cut green before play started. From 12 feet, the True Roller made 73% of its putts. Setting up on the same green after the day's play, the True Roller could make only 30% of its 12-footers! The culprit here is the condition of the green after being trampled by numerous big guys with spikes. A closer examination of what is happening on the greens follows later. From the drastic deterioration of the Westchester greens over a day's play, it is clear that green quality has a significant influence on the percentage of putts made. To be fair, the dramatic drop in green quality occurred on pro-am day, where there is a large volume of play with distracted amateurs, and it was in the era of destructive metal spikes on golf shoes.[5]

If the perfect stroke only produces 30% successes from 12 feet, we cannot get too critical about the pros making only 26% in 1988. To be consistent, though, how much credit can we give Jack Nicklaus for his pressure putt in the 1986 Masters? Maybe, as Nicklaus implicitly acknowledged, he had a little luck.

Engineers Frank Werner and Richard Greig found similar results with their own putting machine. Over a variety of distances, they determined that the standard deviation of the lateral position of the ball is about 2% of the distance of the putt.[6] For our 12-foot putt, 2% is almost 3 inches. Werner and Greig's findings imply that 32% of putts that start out perfectly on-line will be 3 inches or more off-line by the time they reach the hole. Since the radius of the hole is only 2.125 inches, all of those putts will miss. Of the putts that roll exactly 12 feet, almost half will be wide of the hole.

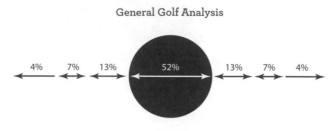

Figure 3.1 Outcomes of a perfectly struck 12-foot putt, with a standard deviation of 2% of distance

Figure 3.1 shows the likelihoods of different left-right positions of a perfectly struck 12-foot putt using the 2% standard deviation rule. Only 52% of the putts stay within the hole, while 26% go up to an inch-and-a-half wide, and another 14% finish up to 3 inches wide of the hole. If you are keeping score, that leaves 8% of the perfectly struck putts more than 3 inches wide of the hole! The news gets worse as the distance increases. From 25 feet, the standard deviation is up to 6 inches, so that about 5% of perfectly struck putts will be more than a foot wide of the hole. Approximately three-fourths of putts struck with the perfect line will stray outside the hole.

Finally, there is one more piece of bad news. Werner and Greig found that distance has the same 2% rule. That is, a putt that should roll exactly 12 feet will actually have a random distance with standard deviation equal to 2% of 12 feet.[7] This is the same standard deviation (about 3 inches) as for the left-right movement of the ball, so that figure 3.1 gives the percentages for a putt struck from left to right with "perfect" speed (that is, a speed that should produce exactly 12 feet of roll). Ignoring the sideways movement of these putts, 24% of these putts stop short of the hole. Another 24% would go past the hole, although, if the line is accurate enough, these would go in.

Never Up, Never In

Of course, the good news in all of this is that you can stand (or sit) tall at the 19th hole. That blown 6-footer on the 18th hole which you swear broke to the left? Just quote the above research: it might not have been your fault. There is a lot of randomness in putting, and putts get bumped significant amounts by small imperfections in the greens.

What should you do about this when you're standing over that 6-footer on the 18th? For starters, relax. Shaking uncontrollably and screaming, "I have no chance. It's all random!" will not help. And, if you are in the group in front of me, please do not try to "read" the footprints. That attention to detail is not attractive in a Ryder Cup competitor, much less a weekend golfer.

The important lesson to take from this is that the perfect speed on a putt is not the speed that allows the ball to die at the hole. As noted above, distance is subject to the random bumps of the green, and your perfect putt could stop inches short of the hole. In other words, "never up, never in" is good practical advice.[8] Of course, it is possible to hit putts too hard, and a firm putt needs to hit the center of the cup. You do not get partial credit for lip-outs.

What is the best speed to hit a putt? Dave Pelz experimented with his True Roller and measured the speed at which the most putts actually went in the hole. In Pelz's tests, a target distance of approximately 17 inches past the hole maximized the likelihood that the putt went in.[9] This does not depend on the length of the putt. Unfortunately, if your target is 17 inches past the hole, sometimes the ball will speed well beyond the hole. If your main desire is to have a simple tap-in for your next putt, a more conservative approach would serve you well. However, in match play, facing a putt to halve the hole, Pelz says that 17 inches past the hole should be your target.

In stroke play, you might not want to be so bold. Instead of maximizing the probability of making the putt, you are probably more interested in minimizing the total number of putts taken. This is the approach of Werner and Greig.[10] Being bold might result in more one-putts, but it is not a good play if it also results in many more three-putts. With this more conservative philosophy, the ideal distance does depend on the length of the putt as well as on the speed of the green and the quality of the putter. Recommended distances range from 5 to 15 inches past the hole for most situations, reaching 17 inches only on fast greens (12 on the Stimp meter[11]) and for narrow distance ranges (around 12 feet for a mediocre putter and 20 feet for a good putter).

One unusual aspect of Werner and Greig's work is that, for a long putt by a high handicapper, the ideal target distance is short of the hole. This certainly contradicts the "never up, never in" motto. However, there is good mathematics behind it. For a weak putter at 50 feet from the hole, the only realistic goal is to lag the first putt close enough to escape with a two-putt. Of course, getting a perfect line is not likely. This presents us with a geometry question. For an off-line putt such as that in figure 3.2, at which point is the ball closest to the hole?

The governing principle is that, at the closest point, the line from the ball to the hole makes a right angle with the ball path. As shown in figure 3.3, this occurs at a point that is slightly short of the hole. This shows that, for a long putt by a mediocre putter, short can be a good play. None of this analysis takes into account the slope of the green. In many cases, you would rather leave yourself a 4-foot uphill putt than a 3-foot sidehill putt. For most putts, however, the best advice is to be confident, aim a few inches past the hole, and do not be surprised if the putt goes a little random on you.

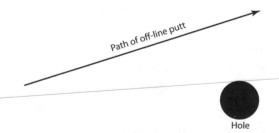
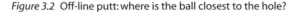

Path of off-line putt

Hole

Figure 3.2 Off-line putt: where is the ball closest to the hole?

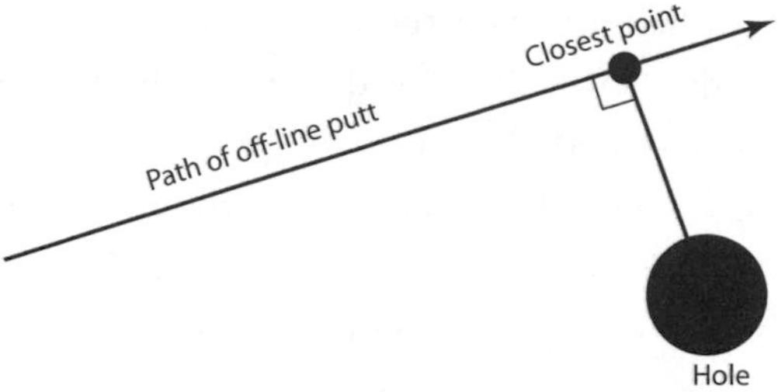

Figure 3.3 Closest point, at right angles with the hole

Sadly, the bad luck does not necessarily end when the ball reaches the hole. Lip-outs are the next topic.

The Unkindest Putt of All

One of the most famous hustles in golf history may have ended with an obscure result from mathematical physics. There are several versions of the story, but Raymond Floyd's goes like this. In 1965, Floyd was well established on tour with two victories. He was living in Dallas, Texas, and ventured out to Tenison Park Golf Course to check out its famous gambling scene. After winning a few bets, Floyd was approached by the legendary hustler Titanic Thompson, who had a big money venture in mind.

Thompson flew Floyd flew out to El Paso to challenge a local hotshot. When Floyd drove out to the course to practice, a young man grabbed his bag and got him set to play. Floyd introduced himself, said he was there to play some guy named Lee Trevino, and asked if the bag man knew him. "That's me," said Trevino. Several thousand dollars, a lot of money in 1965, was wagered on the match. Floyd strutted out and shot 65 ... and lost by two. He was not amused to lose to an unknown who "hit this little old screamer, never gets the ball in the air." So, he went out the next day and shot 64 ... and lost again.[12] Apparently, Thompson desperately needed money and backed Floyd for another round.[13]

After all, Floyd had just won the St. Paul Open and recently finished sixth in the U.S. Open. On the 18th hole, both players had eagle putts to shoot 63. Floyd drained his. In Floyd's words, "I can still see Lee's putt. There's no way it should not have gone in the hole. His ball kind of dives down in the hole, then comes out and sits on the lip." Counting himself lucky and not yet broke, Floyd left town and returned to the "easy" life on the PGA Tour.[14]

What does mathematics have to do with this story? It turns out that mathematicians have analyzed the path of a golf ball after it goes in the hole. In fact, a version of it was used as a question at Cambridge University in the famous Mathematical Tripos examination and has been commented on by J. E. Littlewood, one of the most famous mathematicians of all time.[15] The assumption is that the ball rolls along the side of a cylinder (the cup) under the force of gravity. The rolling assumption is unlikely to be met in golf, since many putts bounce off the lip or drop directly to the bottom of the hole. However, a putt trying to enter the edge of the hole will sometimes roll, and then our assumption is met.

Most people expect that the ball would spiral down to the bottom of the hole. This is partly correct. Viewed from above, the ball does appear to move around its circular boundary at a constant rate. The surprising part of the solution is that the vertical motion of the ball turns out to be alternating periods of downward motion and upward motion.

To solve this problem, two mathematical techniques are used to transform a seemingly impossible problem into a difficult but workable one. The geometry of the ball's path is three-dimensional, so three variables are needed to describe its position. One variable measures the vertical (height) position. The other two variables measure its position as seen from directly above the hole. For these two, the polar coordinates introduced in the previous chapter simplify calculations considerably (see figure 3.4). An obvious

Figure 3.4 Polar coordinates

advantage of polar coordinates is that, for a ball rolling on the cylinder, the value of r is constant.

The other trick is to use a moving reference frame. This means that instead of choosing a particular point as the origin, as indicated in figure 3.4, the origin moves with the ball. Depending on the problem to be solved, it can be more convenient to place the origin at the center of the ball or at the contact point between ball and cylinder.[16] The result is counterintuitive. Both the angle θ and the height z oscillate. This means that the ball alternately spirals down and then spirals up, then down and up again until it loses contact with the side. One of the ways for the ball to lose contact with the side is to spiral back up out of the hole. Figure 3.5 shows a possible path for a ball going down into the cylinder and then coming back out of the hole. Now that's a power lip-out!

A Random Walk Spoiled

Finally, let's take a more detailed look at the mathematics of bad luck on the green. In mathematics, *modeling* refers to the process

Figure 3.5 A lip-out, after the ball
dips in to the hole

of approximating some aspect of reality with a mathematical description. The construction of a good model depends on in-depth
knowledge of the process being described mixed with reasonable
guesses and simplifications. Like a scientific theory, a model is used
to make predictions. If the predictions turn out to be highly inaccurate, then the model must be abandoned. Good predictions count
mostly as circumstantial evidence of the model's usefulness, not as
proof that the model is correct.

We will look at two models of random bumps on the greens.
When I have mentioned the variability of greens to colleagues, their
speculations have included spike marks, loose dirt, the dimples on
the ball, and the wrath of the gods (this last one, obviously, from
one who plays a lot of golf). Dave Pelz has written extensively about
footprints on the green being to blame.[17] Some modeling will help
us evaluate these ideas.

Model 1 is based on the idea that the ball is constantly buffeted
by spike marks, dirt, stiff blades of grass, etc. Mathematicians refer
to this as a "random walk" (also called a "drunkard's walk," which
would give a different meaning to the reference to golf being "a
good walk spoiled"[18]). Imagine someone who is generally walk-
ing straight ahead but whose every step tips slightly to the left or
to the right, with left or right being equally likely on each step.
The actual path that the person walks might look like figure 3.6,
where the person's first step leans to the left, the second step

tips back to the right, the next two steps lean to the left, and so on.

We can model a putt under constant bombardment as a random walk with infinitesimally small step size. That is, take the path in figure 3.6, extend it for a very large number of steps outside the box, and then shrink it back to fit in the box. Repeat this process several times. You now have a path with so many bends that it is impossible to distinguish individual steps. An example of such a path is shown in figure 3.7.

Properly scaled, this is considered a very good model for Brownian motion, the random movement of tiny particles suspended in a fluid. The math in this model is quite elegant and historically important, as Albert Einstein himself did some of the early work in the analysis of Brownian motion.[19] For our purposes, it is also verifiably wrong. The standard deviation of the lateral displacement of a Brownian motion path is proportional to the *square root* of distance. Werner and Greig found that the standard deviation of golf putts

Figure 3.6 First six steps of a random walk from left to right: the first step tips to the walker's left, the second back to the right, the next two to the left, and so on.

Figure 3.7 Brownian motion

is proportional to the distance of the putt (the 2% rule mentioned earlier). Going from a 5-foot putt to a 20-foot putt, the standard deviation of golf putts increases by a factor of 4. The Brownian motion model has a standard deviation that increases by a factor of $\sqrt{4} = 2$. So, model 1 fails.

A Good-looking Model

Model 2 is based on the idea that the ball must travel through footprints on the way to the hole. These indentations change the angle at which the ball moves. (By contrast, the ball in the random walk model is always aimed straight ahead, with random left/right bumps added.) To help define the variables used, figure 3.8 shows a putt of d feet that has been bumped x feet off-line and is moving at an angle A to the straight path.

We next look at the ball after it has moved forward a distance y feet along the angle A. At this point, it may or may not hit a footprint and deflect. To be precise, with some probability p its angle is altered, and with probability $1 - p$ its angle remains equal to A. If it does hit a footprint, then half the time the angle increases to $A + b$, and half the time the angle decreases to $A - b$. This is illustrated in figure 3.9.

We continue this process until the ball has traveled the full distance d and then measure the amount x that the ball is off-line. The angular deflection b can be chosen so that the standard deviation of x is two percent of d.[20] This matches both the experimental data of Werner and Grieg and the physical mechanism described by Pelz. Figure 3.10 shows the path of a putt generated by model 2. While the path may look smooth and resembles a putt that breaks to the right, it is actually produced by model 2, a putt on a flat green

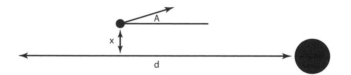

Figure 3.8 The ball is x feet off-line, moving at an angle of A from the straight path

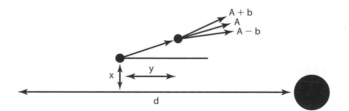

Figure 3.9 After moving y feet along angle A, the ball's angle may change from A to A + b or A − b

Figure 3.10 A putt deflected by footprints

which is struck straight at the hole and then deflected by a series of footprints.

Model 2 matches one aspect of the experimental data, so it remains a possible explanation of how the irregularities of a green affect a putt. This model has the advantage of supporting Dave Pelz's explanation that footprints leave the green a "lumpy" carpet that alters the trajectory of the putt as it travels toward the hole.[21]

The Back Tee: Modeling Details

To close the chapter, I want to fill in some of the details of model 2 above. If the mathematical "lie" becomes unplayable, you can safely pick up and "play through" to the next chapter.

As illustrated in figure 3.9, x measures how far off-line the ball has drifted. If the ball moves a horizontal distance of y at an angle of A, the change in x is $y \tan A$. The total deflection x is thus

$$x = y \tan A_1 + y \tan A_2 + \ldots y \tan A_n,$$

where A_1 is the angle for the first step, A_2 is the angle for the second step, and so on for all n steps, assuming that each step is of equal length y. The angles can change by $\pm b$ at each step, but they do not have to change at all. In mathematics, a surprising amount of progress can be made just by giving names to quantities. I will denote the changes in the angle by c_1, c_2, and so on. If the putt starts out on-line, then the initial angle is 0 and $A_1 = c_1$, $A_2 = A_1 + c_2 = c_1 + c_2$, and so on.

For "small" angles, $\tan A \approx A$. This approximation makes the remainder of the analysis much easier to follow (but is not needed to get a result). Then

$$x \approx y[A_1 + A_2 + \ldots + A_n]$$

$$= y[c_1 + (c_1 + c_2) + (c_1 + c_2 + c_3) + \ldots + (c_1 + c_2 + \ldots + c_n)]$$

$$= y[nc_1 + (n-1)c_2 + (n-2)c_3 + \ldots + c_n].$$

We are interested in the standard deviation of x. The mean value of x is 0, which makes the standard deviation equal to $\sqrt{E(x^2)}$, where $E(x^2)$ is the expected value or mean of x^2. The randomness in our final formula for x resides in the changes c of the angle. I assume that the c's are independent but not identically distributed. Each c is 0 with probability $1 - p$ and $\pm b$ with equal probabilities $\frac{1}{2}p$. However, the probability p is not constant. Thus, the c's have mean 0 but different variances.

In particular, I assume that the probability of a deflection depends on the distance from the hole. The closer to the hole the ball is, the higher the probability of a deflection (the larger p is). The logic behind this assumption is that footprints will tend to pile up near the hole and be relatively sparse far from the hole. For a putt of any distance, assume that the p-value for the last y feet is some value q. This means that the probability p_n associated with c_n equals q. Assuming that the probability is inversely proportional to distance from the hole, the probabilty associated with c_{n-1} is $p_{n-1} = \frac{1}{2}q$, the probability associated with c_{n-2} is $p_{n-2} = \frac{1}{3}q$, and so on. This means that $E(c_n^2) = b^2 q$, $E(c_{n-1}^2) = b^2 \frac{1}{2}q$, and so on. In terms of x, we get

$$E(x^2) = y^2[n^2 E(c_1^2) + (n-1)^2 E(c_2^2) + \ldots + 4E(c_{n-1}^2) + E(c_n^2)]$$

$$= y^2[n^2 b^2 \frac{1}{n}q + (n-1)^2 b^2 \frac{1}{n-1}q + \ldots + 4b^2 \frac{1}{2}q]$$

$$= y^2 b^2 q[n + (n-1) + \ldots + 2 + 1]$$

$$= y^2 b^2 qn(n+1)/2$$

$$\approx y^2 n^2 b^2 q/2.$$

The total distance of the putt equals yn, so the standard deviation of x is given by

$$SD = \sqrt{y^2 n^2 b^2 q/2} = yn(b\sqrt{q/2}),$$

which equals 2% of the distance if $b\sqrt{q/2} = 0.02$. The amount of deflection b is therefore connected to the probability q of a deflection near the hole. We can think of the parameters b and q as measures of the quality of the green.

FOUR

The Rivalry
Cautious and Risky Strategies

I could have played it safe,
but that wouldn't have been me.

—*Arnold Palmer*

Rivalries play a critical role in defining us as sports fans. Do you root for the Tar Heels or the Blue Devils?[1] Red Sox or Yankees? Your answers may determine how loud your next bar conversation gets.

In golf, the rooting process is highly personal. We can see our heroes scowl and laugh as they respond to tragedies and triumphs. If you are a Tiger fan, it is probably because you admire his dedication and fierce competitiveness and want to see history made. If you are a Phil fan, you probably enjoy his fearlessness and joyful competitiveness and want to be thrilled.

The sport of golf was blessed with the ultimate rivalry in the 1960s when sports television came of age. In many ways, Arnold Palmer and Jack Nicklaus established the characters that Phil and Tiger played in the 2000s. Palmer was the gambler, winning and losing tournaments with late charges and collapses that turned staid, old golf into high drama. Nicklaus was the calculating perfectionist who rewrote the record book. Since Palmer established himself first, he was "the King" and everybody's favorite. Nicklaus did not just serve as Palmer's biggest rival, he was the usurper who dethroned the King. The public saw them as wrestling hero and villain, and large galleries rooted for Palmer with a fervor that was

only partially moderated by the remarkable class and sportsman-ship displayed by both men. These competitors truly wanted to beat each other in every way, but there was also respect, admiration, and even envy between them which created a special bond.[2]

This chapter analyzes a mathematical caricature of the Jack/ Arnie and Tiger/Phil rivalries. In a variety of settings, we will explore which of two golfers of "equal" ability will have an advantage. One player is more consistent and the other is more erratic. Alternatively, you can think of them as being cautious and risky golfers, respectively.

The mathematics of this study has a surprising history. The great G. H. Hardy, a preeminent mathematician of the early 1900s, first defined and analyzed the "characters" in this study. The irony is that Hardy and his colleague J. E. Littlewood were strong proponents of pure mathematics of the most cerebral kind, yet both made contri-butions to the mathematics of golf (see chapter 3 for Littlewood's analysis of lip-outs).

Hardy Golf

Hardy's golfer is capable of exactly three types of shots: excellent, normal, and poor. An excellent shot E reduces the player's score on a hole by one; a poor shot P increases the player's score on a hole by one; and a normal shot N maintains the golfer on a par pace. For the sake of simplicity, every hole will be a par 4.[3] Then, a shot sequence of NNNN finishes the hole in par 4. However, a sequence of NEN represents a score of 3, with the excellent shot reducing the score from a par 4 to a birdie 3. A shot sequence of NPNNN represents a bogey 5, with the poor shot increasing the score by one.

There is some ambiguity in this description, and the interpretation makes a large difference mathematically. The sequence NEE could be thought of as a birdie because there are three shots listed or an eagle because the second excellent shot might reduce the score by a second stroke. We will count strokes, so this is a 3. The second excellent shot is, in a sense, wasted because it produces the same outcome that a normal shot would. The situation is analogous to stroking a birdie putt with perfect speed and line. If the putt is from 3 feet away, the perfect putt is somewhat wasted because a normal, slightly imperfect putt would also have gone into the hole.

Each shot is assumed to be random and independent. That is, the chance of an excellent shot is the same whether the previous shot was excellent, normal, or poor. (Don't you wish this were true in real life?) The probability of an excellent shot is some number p and the probability of a poor shot is the *same* number p. Different Hardy golfers may have different p-values, but if a shot is not normal, it has the same chance of being excellent as it does of being poor. In this sense, every Hardy golfer has the same skill level.

The "rivalry" for this chapter is between a Hardy golfer with a small p-value and one with a larger p-value. For clarity, I will use $p = 0.1$ and $p = 0.2$. The Hardy golfer with $p = 0.1$ hits on the average 1 excellent shot, 1 poor shot, and 8 normal shots for every 10 shots. This is C-golfer, our consistent, cautious golfer. The Hardy golfer with $p = 0.2$ hits on the average 2 excellent shots, 2 poor shots, and only 6 normal shots for every 10 shots. This is

R-golfer, our erratic, risky golfer. The question is whether these golfers really are equal. Is there an advantage to being consistent, or is it better to have more frequent flashes of brilliance? In real life, Jack and Tiger won more often than Arnie and Phil. We will see how their mathematical counterparts perform.

A Mean Start

A good place to start is to compute the average score for a round. While this does not tell us the odds of either player winning a match, it gives us a useful benchmark. More importantly for a mathematician, the calculation turns out to be a very elegant result.[*] A detailed description of this calculation is deferred to the end of the chapter.

The result is that on a par 4 the average score for the consistent golfer ($p = 0.1$) is almost exactly 4.1, and the average score for the erratic golfer ($p = 0.2$) is almost exactly 4.2.[4] The higher the p-value, the more erratic the golfer is and the higher the average score will be. In general, the average score on a hole for a Hardy golfer is approximately p over par, a wonderfully simple result. Multiplying by 18 holes, C-golfer averages 73.8 to R-golfer's 75.6.

There is a (nearly) two-stroke advantage for the consistent golfer. The numerous calculations to follow will expand on this result, but there is an immediate lesson to be learned here. If great and horrible shots are equally likely, your average score will improve if you play it safe and hit as many normal shots as possible. In other words, go for the lowest p-value you can. It may not be as much fun to keep hitting safe, basic shots, but it does turn out to be more effective.

---

[*]An important criterion of any non-trivial mathematical result is that it be "elegant." The use of that word by mathematicians may surprise you, but it is an extremely common word choice for describing mathematical work. "Elegant" implies both beautiful and simple, and the world of mathematics is all about finding the simplest descriptions of connections between mathematical entities.

The average score does not fully answer the question of who will win a match. The answer to that question depends in part on what type of match is played.

Stroke Play

In stroke play, both players go 18 holes, counting all strokes, and the lower total wins. As noted above, C-golfer averages almost 2 shots fewer than R-golfer for a round. The real question, however, is what the odds are for each player to win a single match.

A comparison of average scores does not necessarily answer the question. If R-golfer averages 75 by shooting 70, 70, 70, and 90, that might be good enough to win 3 out of 4 matches. To determine the odds of winning the match, it is necessary to compute the probability that each player shoots 70, 71, 72, and so on and compare the probabilities.

The actual probabilities are given in table 4.1. Figure 4.1 shows the probabilities graphically for each golfer.

The erratic R-golfer is more likely than C-golfer to shoot a historic 59 (or any score of 67 or less). This is balanced by the fact that R-golfer is also more likely to blow up to 77 (or any higher score). The bottom line is that C-golfer wins about 56.8% of the rounds, R-golfer wins about 37.1%, and about 6.1% are tied. To illustrate this result, I had my computer simulate some rounds using the Hardy model. The following group of 20 rounds each is representative of the other simulations, and the averages match the theoretical averages closely.

C-golfer: 71, 77, 71, 74, 74, 76, 79, 73, 73, 74, 74, 71, 74, 73, 74, 78, 73, 69, 77, 74

R-golfer: 80, 66, 79, 75, 78, 89, 68, 76, 72, 72, 80, 74, 72, 82, 78, 75, 74, 69, 77, 71

Notice that C-golfer wins 11 rounds (55%) and ties 2 (10%). R-golfer turns in some wildly erratic rounds, ranging from a hot

Table 4.1 Probabilities for shooting scores from 59 to 86, rounded to 4 decimals, for C-golfer and R-golfer

Score	C-golfer (p = 0.1)	R-golfer (p = 0.2)
59	0	0.0002
60	0	0.0005
61	0.001	0.0011
62	0.0003	0.0020
63	0.0009	0.0036
64	0.0022	0.0061
65	0.0049	0.0098
66	0.0101	0.0148
67	0.0188	0.0213
68	0.0315	0.0290
69	0.0482	0.0378
70	0.0672	0.0471
71	0.0857	0.0562
72	0.1005	0.0642
73	0.1086	0.0705
74	0.1085	0.0745
75	0.1007	0.0759
76	0.0870	0.0745
77	0.0703	0.0706
78	0.0532	0.0648
79	0.0380	0.0575
80	0.0255	0.0494
81	0.0162	0.0412
82	0.0098	0.0333
83	0.0056	0.0262
84	0.0031	0.0200
85	0.0016	0.0149
86	0.0008	0.0107

66 to an embarrassing 89, while winning 7 matches (35%). The consistent player has a definite advantage. (In the next chapter, we will see that the USGA handicap system is not designed to even out this match.) The lesson is that, even if the odds are 50-50 of pulling off a risky shot, the conservative shot is better in stroke play.

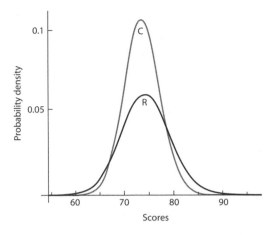

Figure 4.1 Probabilities of 18-hole scores for C-golfer and R-golfer

Match Play

In match play, each hole is a separate contest, and the player who wins the most holes wins the match. The standard thinking is that match play is kinder to erratic golfers. A terrible shot leading to a "cane" (7) or "snowman" (8) can put you several strokes behind in stroke play, but only costs you one hole in match play. Meanwhile, the erratic player has a higher propensity for making birdies, which can win holes. To compare C-golfer and R-golfer in match play, you need the probabilities for each golfer to make a 2, 3, 4, and so on for a single hole. The probabilities for one hole can then be manipulated to give the probabilities for an 18-hole match. The probabilities, rounded to two decimal places, are:

	2	3	4	5	6	7	8
C-golfer	0.01	0.21	0.52	0.20	0.05	0.01	0.0
R-golfer	0.04	0.28	0.34	0.20	0.09	0.03	0.02

The table shows that the erratic R-golfer is more likely to make eagle or birdie but is also more likely to make double bogey or worse. The consistent C-golfer pars more than half of the holes.

Comparing probabilities, it turns out that C-golfer wins 36.3% of the holes, R-golfer wins 35.8% of the holes, and 27.9% of the holes are halved. Again, there is an advantage for the consistent player, although it is much smaller this time. Extrapolating over an 18-hole match, C-golfer will win 45.5% of the matches, R-golfer will win 43.3% of the matches, and 11.2% of the matches will be draws. Match play is, indeed, kinder to the erratic player, but the consistent player still has the edge. The conclusion is that, in the standard forms of head-to-head competition, the more consistent you are, the better. If you insist on going for the impossible shot, it will cost you less in match play than in stroke play.

Best Ball

Even though an erratic player is likely to lose to a more consistent player one-on-one, perhaps two erratic players working together can beat two consistent players in "best ball." In this type of match, each player plays each hole to completion. The team's score on the hole is the better of the scores of the two players on the team. So, the lesser-used term "better ball" is a more accurate name.

The most extreme example I have experienced of the oddities of team play came in a best ball championship match. My partner and I both started the round playing horribly. Each of our individual front-nine scores would have been well into the 40s; however, on each hole one or the other of us would make a putt, so our best ball score was even par.[5] Our opponents each shot par individually but missed numerous short birdie putts and were only 1-up at the turn. Aside from casting some confused and increasingly irritated glances at us, they kept their cool and were rewarded on the back nine. My partner and I each had better individual scores on the back nine than did our opponents, but on this nine they made all of the clutch putts and closed us out. On both nines, the team scores were very different from the individual scores.

The mathematical question is how the erratic R-golfer fares in best ball competition. In individual play, his relatively high chance of making a hole-winning birdie is more than offset by numerous hole-losing bogeys and doubles. With a partner to save him on his bad holes, perhaps he will finally be able to beat the consistent C-golfer.

One possible match is between two erratic players and two consistent players. To analyze this match, the probabilities of the two R-golfers making a 2, 3, 4, and so on are combined to find the probabilities for the team's score. Details on how this can be done are at the end of the chapter. The calculation is repeated for the team of two C-golfers. The results are shown below.

	2	3	4	5	6
Team R	.08	.46	.34	.10	.02
Team C	.02	.37	.54	.06	.01

Team R is much more likely to make birdie or eagle, but is this enough to offset the 37% birdies and 54% pars of Team C? It turns out that, with these probabilities, Team R wins 38% of the holes, Team C wins 26%, and 36% of the holes are halved. Score one for the risk-takers.

Another possibility is to pair an erratic player with a consistent player. The logic is that the consistent player removes the chance of the team making bogey or worse, while the erratic player makes enough birdies to win the match. How does this play mathematically? The team breakdown is as follows:

	2	3	4	5	6
Team RC	.05	.42	.44	.08	.01

In a match against Team C (two C-golfers), Team RC wins 33% of the holes to Team C's 27%, with 40% halved. In a match against Team R (two R-golfers), Team RC wins 29% of the holes to Team R's 35%, with 36% halved.

It seems that, in best ball, the more erratic players you have, the better. Said differently, if you can balance your great shots and bad shots, it pays in best ball to be aggressive and take chances. Chris Conklin approached the issue of team play as a college golfer. In a college team match, five golfers play for each team, with the team score being the sum of the best four 18-hole scores. Conklin's hypothesis was that supplementing three consistent players with two erratic players might produce a better team score than using five consistent players. Using players with $p = 0.1$ and $p = 0.05$, his simulations showed that the team with the most consistent players always had an advantage.[6]

Skins

The ultimate game for aggressive golfers is "skins." In this game, all four players record their scores on a hole and the lowest score wins. If two or more players tie with the lowest score, then the group moves on to the next hole. The phrase "two tie, all tie" means that everybody is eligible to win the next hole, regardless of how they played on the previous hole. Many golfers love to wager, and there are several ways to bet on skins. The most common is to "carry over" all bets. If the bet is $1 per hole and the first hole is tied, then everybody plays the second hole for $2. If the second hole is also tied, then everybody plays the third hole for $3. An erratic golfer can have a string of bad holes and still be in line to win all of the money if the other three are cooperating and tying every hole.

Who has the advantage if the foursome consists of three C-golfers ($p = 0.1$) and one R-golfer ($p = 0.2$)? The calculations show that, on a given hole, each of the C-golfers has a 10.8% chance of winning the hole, the R-golfer has a 17.7% chance of winning the hole, and 49.9% of the holes are tied. The erratic golfer has a clear advantage here.

If the foursome consists of two C-golfers and two R-golfers, the percentages change only slightly. On a given hole, each of the C-golfers has a 10% chance of winning the hole, each of the R-golfers has a 16% chance of winning the hole, and 48% of the holes are tied. Again, about half of the skins are carried over.

The number of holes halved does not vary much with the composition of the foursome, at least for foursomes composed of C-golfers and R-golfers. If all four golfers are C-golfers, then 53% of the holes are tied. If all four golfers are R-golfers, the percentage of ties drops, but only to 45%.

Tournaments

Now let's place several C-golfers and R-golfers into a tournament setting. Recall that, for a single round, C-golfers average almost 2 strokes better than R-golfers. Nevertheless, R-golfers have a greater chance of scoring 67 or lower. Generally, C-golfers will perform better, but our question here is whether there will always be at least one "hot" R-golfer to take over the tournament. This analysis is done for a 140-player tournament with 90 C-golfers and 50 R-golfers.

In a one-round tournament, an R-golfer wins 56% of the time, a C-golfer wins 29% of the time, and R-golfers and C-golfers tie for first 15% of the time. Even though the R-golfers are outnumbered, the odds are high that one of them will go low and win the tournament. Of course, many of the R-golfers will blow up and be at the bottom of the leaderboard.

This represents an interesting statistical "paradox." For an individual golfer, the lower your p-value is, the better your average will be. However, for a one-round tournament, the absolute best score is likely to come from a golfer with a higher p-value. If the goal is to improve your chances of winning without worrying about your overall placement, try some risky shots. If you would like a nice top-20 finish, conservative plays are best.

The situation changes for a four-round tournament. While a gambling R-golfer might steal the lead for one round, the odds that the gambles continue to pay off decrease as the tournament progresses. In a four-round tournament, a C-golfer will win 68% of the time, while R-golfers win 29% of the time. For the pros, then, a long-term plan is important, and conservative play is often rewarded.

This points out an important difference between an event like the U.S. Open and an event like the Ryder Cup. Long-term strategies that lead to good finishes in the Open do not always translate to victories in the one-round sprints that comprise the Ryder Cup. This is true mathematically, as we have seen with our C-golfers and R-golfers, and psychologically, as the Ryder Cup teams show every two years.

The Hardy Open

To illustrate the above results, I simulated a four-round tournament consisting of 90 C-golfers and 50 R-golfers. I call it the Hardy Open, since it consists of Hardy golfers but shows some of the oddities often found in Open championships.

The first-round lead is grabbed by two unknown R-golfers who shoot 65, R21 and R43 (the 21st and 43rd R-golfers in the field). Third place at 66 and fourth place at 67 are also taken by R-golfers.

R21	65
R43	65
R7	66
R27	67
C18	68
C77	68
R8	68
R48	68

The pressure of the overnight lead gets to R43, who blows up to an 85 and misses the cut. R21 fires a second-round 70 to grab the halfway lead at 135. The cut of the top 70 plus ties falls at 148. Of the 90 C-golfers, 56 make the cut. Only 20 of the 50 R-golfers make the cut. A course-record 61 is turned in by R12. A trio of R-golfers sit at 139, with four C-golfers at 140 and 13 of the next 15 spots taken by C-golfers.

R21	135
C14	137
R12	137
R49	137
C49	138
R7	139
R24	139
R27	139

On "moving day" (third round), R21 and C14 move out of contention with an 80 and 81, respectively. R12 is unable to duplicate the magic of his 61, falling to a 78. He will, however, rebound in the final round with a 68 to finish tied for 6th. C73 grabs a share of the lead with a 67, tying the unusually consistent R49's 69-68-71.

C73	208
R49	208
C53	209
R24	209
C51	210
C85	210
R37	210
C77	212

In the final round, C77's strong 67 wins the tournament. He receives the traditional laurel and a Hardy handshake for the winner. Third-round coleader R49 cannot handle the pressure of being in the final group and shoots 83. Meanwhile, second-round leader R21 recovers from a third-round 80 to nearly steal the tournament with a final round 65. In the broadcast booth, Johnny Miller notes that the 63 he shot at Oakmont would have won the tournament.

C77	279
R21	280
C73	281
R24	281
R26	281
C25	283
R12	283
C59	284
R37	284

Of the 12 players in the top-10-plus ties, half are C-golfers and half are R-golfers. The next 12 players are all C-golfers. Of the 54 players in the top-50-plus ties, 40 are C-golfers. The conclusion is that playing risky golf gives you a shot at the top spot, but cautious play leads to made cuts and good money.

Although I described the tournament as if players were responding well or poorly to pressure, remember that these were all results that were randomly generated on the computer. This is not to say that the golfers we watch in tournaments are not choking, but it is instructive to notice that random golfers show some of the same scoring characteristics. In particular, the phenomenon of the unknown first-round leader was present here. The fast exit of R43 was not due to pressure but to some bad random numbers. As well, R27 followed a first-round 67 with rounds of 72-80-76. Notice that his scores in the last three rounds averaged 76, which is the average score for an R-golfer. This is called *regression to the mean,* where an unusually good (lucky) round is followed by a more typical, not-so-good round. I suspect that this explains the collapse of many first-round leaders.

Great Expectations

Hardy's model gives us some insight into the question of how advisable risky shots can be. In this section, actual data from PGA tournaments gives us a different way to approach the same question. The mathematical tool that we will use is the *expected value*, a weighted average that can give us information about whether cautious strategies pay off in the long run.

To illustrate the expected value, let's start on the green. Suppose that we one-putt 30% of the time and two-putt the other 70% of the time. This cannot literally hold for an 18-hole round, because 30% of 18 is $0.3 * 18 = 5.4$, and one-putting 5.4 times is not possible. However, over a large number of rounds it would be possible to have these averages, in which case we would average 5.4 one-putts and 12.6 two-putts per round, for an average (expected value) of $5.4(1) + 12.6(2) = 30.6$ putts per round.

We will now apply expected values to the question of whether to unleash the driver on a short par 4. Let's say that our hole is 380 yards long, and our options are to bomb a driver 320 yards or smooth a 240-yard hybrid into the fairway. In addition, assume that we hit the fairway 90% of the time with the hybrid and 40% of the time with the driver. Figure 4.2 shows average scores on par 4s on the 2008 PGA Tour, when playing from different distances from the fairway compared to the rough.

An interesting fact illustrated in figure 4.2 is that missing the fairway costs the pros about one-fourth of a shot on a par 4. That is, for a given distance from the hole, the average score is about 0.25 strokes higher from the rough than it is from the fairway.

Using the numbers behind figure 4.2, we can evaluate strategies. Taking the safe hybrid from the tee leaves us 140 yards out. We are in the fairway 90% of the time, from where we expect to average 3.883 strokes on the hole. We are in the rough 10% of the time, from where we average 4.180 strokes. The expected value for

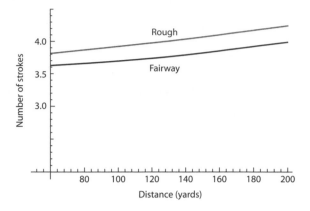

Figure 4.2 Average scores on 2008 PGA Tour on par 4s with second shot from the given distance and position

the safe play is then $0.9(3.883) + 0.1(4.180) = 3.9127$. Taking the driver from the tee, we leave ourselves 60 yards from the hole. In the fairway, we average 3.685 strokes, and in the rough we average 3.918 strokes. The expected value if we hit 40% of the fairways is $0.4(3.685) + 0.6(3.918) = 3.8248$. Comparing the expected values, we see that the driver gives us a lower average score, saving us about 0.1 strokes on the average.

Of course, this analysis is only as good as the data we are given in figure 4.2. If the particular hole that we are on has a lake covering most of the last 100 yards to the green, then the averages in figure 4.2 may not apply, and we might prefer the cautious tee shot.

What if we do not hit the fairway with our driver 40% of the time? Is the driver still the better play if we only hit 30% of the fairways? Or 20%? This is a case in which the mathematician's use of variables can save a substantial amount of work. Instead of repeating the above calculation for many different percentages, we can answer all of the questions at once by naming the percentage of fairways hit: I'll call it p. The calculations are easier if we have p represent the proportion of fairways hit (so $p = 0.2$ instead of $p = 20\%$).

The hybrid still has an expected value of 3.9127. To compute the expected value for the driver, we use the same format as above, just replacing the proportions 0.4 and 0.6. Notice that these proportions add up to 1, or 100%. If our fairway proportion is 0.4, the rough proportion is $1 - 0.4 = 0.6$. In general, if the fairway proportion is p, the rough proportion is $1 - p$. Then the expected value is $p(3.685) + (1 - p)(3.918) = 3.918 - 0.223p$. For the driver to be the better play, we need $3.918 - 0.223p < 3.9127$. Some algebra shows that this is true as long as $p > 0.0227$. In other words, if we can hit the fairway more than 2% of the time, we should use the driver!

Now you can see why the modern game emphasizes length over accuracy.

I Got Game Theory

The mathematical field of *game theory* gives us another way to assess risk and reward situations. Game theory is often applied to competitions between two players, each of whom must choose among several strategies. In golf, the competition is usually between golfer and course, or golfer versus nature, as in the case of playing in a gusty wind.

On a par 3, the golfer must decide whether to aim for the pin or play it safe. This time, I will make up some average scores. Playing it safe, let's assume that the golfer averages 2.9 if the wind does not blow and 3.1 if the wind does kick up. Aiming for the pin, the golfer averages 2.4 if the wind does not blow and 3.4 if it does. These are summarized in a payoff matrix, shown in figure 4.3.

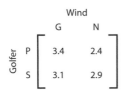

Figure 4.3 Average scores if the wind is gusty (G) or not (N) and if the golfer aims for the pin (P) or plays it safe (S)

One way to analyze this game is to assume that the wind is an active player in the game. This requires a certain amount of paranoia, but I believe most golfers will be able to relate. Regardless of which way the golfer plays it, the wind can force a higher score by choosing strategy G (gusting). Betting on the wind playing this strategy, the golfer would need to play it safe by using strategy S. The golfer accepts an average score of 3.1 and moves on to the next hole.

Assuming that the wind is more of a random phenomenon and less of an evil adversary, we can simply assume that the wind blows with probability p and use expected values as before. If the golfer is going for the pin (strategy P), the expected value is $3.4p + 2.4(1-p)$. This simplifies to $2.4 + p$. If the golfer is playing it safe (strategy S), the expected value is $3.1p + 2.9(1-p)$, which simplifies to $2.9 + 0.2p$. The break-even point is where these two expected values are equal, so that $2.4 + p = 2.9 + 0.2p$. Solving this equation, we get $p = 5/8$. This tells us that, if $p > 5/8$ (that is, we expect the wind to kick up more than 5 out of 8 times), we should play it safe. If $p < 5/8$, we should go for the pin.

In this case, "we should go for the pin" means that the long-run average score for playing this hole under the same conditions numerous times will be better if we aim for the pin each time than if we play it safe each time. The fact is that, as we stand on a tee box, we will play the hole only once. A good long-run average is of little consolation if the wind kicks up this one time and blows our shot into a lake.

A story related in Kevin Cook's *Driven* illustrates this dilemma perfectly. Isabelle Lendl, then eleven years old, was given an assignment by her dad, Ivan Lendl (the tennis star). She was to estimate her average score if she tried to hit a long 3-wood over a lake versus laying up in front of the lake. Figuring that she could carry the lake 20 times out of 100, going for it produced an average score of 5.2 compared to an average lay-up score of 4.8. As fate would have it, that weekend Isabelle faced that exact choice. She went for the green and dumped her shot in the lake. Afterwards, Dad said, "Isabelle, what were you thinking? You told me you could hit that shot only twenty times out of a hundred." Isabelle's response epitomizes the R-golfer in us all: "I thought it was gonna be one of those twenty times."[7]

The Back Tee: Summing Up

To close the chapter, I fill in some of the details left out of the Hardy discussion. If the mathematics here gets a little heavy, feel free to "pick up" and move on to the next chapter.

To analyze each of the games between C-golfers and R-golfers, you need to be able to take probabilities for individual golfers and compute probabilities for the best score in a group. This is an example of what is called an *order statistic*. For example, given four golfers on a particular hole, what will be the minimum score posted by any of the four? In many cases, we are especially interested in *which* golfer has the best score.

To simplify the calculations, suppose that our golfers are capable of five scores on a hole, with probabilities as shown.

	2	3	4	5	6
C-golfer	0	0.2	0.6	0.2	0
R-golfer	0.1	0.2	0.3	0.3	0.1

If two of these golfers are paired together for a best ball match, one question is how to compute the probabilities for the team score. Start with two C-golfers. Since neither player is capable of a 2, the team has 0 probability of a 2. For the team to score a 3, either both players make 3s or one makes a 3 and the other a higher score. The probability that they both make a 3 is the product $0.2 \times 0.2 = 0.04$. Next, consider the possibility that the first C-golfer makes a 3 (probability 0.2), and the second C-golfer makes a higher score. The second player could make a 4 or 5, with probability $0.6 + 0.2 = 0.8$. So, the probability that the first C-golfer makes 3 and the second C-golfer makes higher than 3 is $0.2 \times 0.8 = 0.16$. The other way for the team to make a 3 is for the first player to make higher than 3 (probability $0.6 + 0.2 = 0.8$) and the second player to make a 3 (probability 0.2), which again has probability $0.8 \times 0.2 = 0.16$. Putting this together, the probability of a team 3 is

$$\Pr(3) = 0.2 \times 0.2 + 2 \times 0.2 \times (0.6 + 0.2) = 0.36.$$

Similarly, the team makes 4 if both players make 4s or if one player makes 4 and the other makes higher. The probability is

$$\Pr(4) = 0.6 \times 0.6 + 2 \times 0.6 \times 0.2 = 0.6.$$

Finally, the team makes 5 only if both players make 5, which happens with probability

$$\Pr(5) = 0.2 \times 0.2 = 0.04.$$

Notice that the three probabilities add up to 1, or 100%.

The probabilities for a team of two R-golfers are computed as

$$\Pr(2) = 0.1 \times 0.1 + 2 \times 0.1 \times (0.2 + 0.3 + 0.3 + 0.1) = 0.19$$

$$\Pr(3) = 0.2 \times 0.2 + 2 \times 0.2 \times (0.3 + 0.3 + 0.1) = 0.32$$

$$\Pr(4) = 0.3 \times 0.3 + 2 \times 0.3 \times (0.3 + 0.1) = 0.33$$

$$\Pr(5) = 0.3 \times 0.3 + 2 \times 0.3 \times 0.1 = 0.15$$

$$\Pr(6) = 0.1 \times 0.1 = 0.01.$$

For a match between Team C and Team R, then, the team probabilities are

	2	3	4	5	6
Team C	0	0.36	0.6	0.04	0
Team R	0.19	0.32	0.33	0.15	0.01

The next question might be to compute the probability that Team C wins a hole. There are three possibilities for this outcome. Team C could make 3 while Team R makes 4 or higher. The probabilities are 0.36 and 0.33 + 0.15 + 0.01 = 0.49, respectively. The probability of both occurring is the product $0.36 \times 0.49 = 0.1764$. Team C could make 4 while Team R makes 5 or higher. The probability is $0.6 \times (0.15 + 0.01) = 0.096$. Finally, Team C could win with a 5 if Team R makes 6. This happens with probability $0.04 \times 0.01 = 0.0004$. The total probability of Team C winning is $0.1764 + 0.096 + 0.0004 = 0.2728$, which is slightly larger than 27%.

Team R can win a hole in three ways, as shown.

Team R	Team C	Team R prob	Team C prob	Prob
2	3, 4, 5	0.19	0.36 + 0.6 + 0.04	0.19
3	4, 5	0.32	0.6 + 0.04	0.2048
4	5	0.33	0.04	0.0132

Adding the probabilities in the right-hand column together, the probability that Team R wins is 0.408, or about 41%.

Finally, the teams can tie in three ways, as shown.

Team R	Team C	Team R prob	Team C prob	Prob
3	3	0.32	0.36	0.1152
4	4	0.33	0.6	0.198
5	5	0.15	0.04	0.006

The probabilities add up to 0.3192, for a 32% chance of halving the hole.

Other statistics can be computed in similar ways. The first step is to identify all of the different ways that the event of interest (such as Team C getting a 3 or Team C winning with a 3) can happen. For each possibility, identify the relevant probabilities. Add the probabilities when any of the options are valid (such as Team R making a 4 *or* a 5). Multiply the probabilities when all options have to occur (such as Team C making a 3 *and* Team R making a 3). Then add up all of the component probabilities. When there are more than four or five possibilities, it is very nice to have a computer to do the arithmetic for you. The principles remain the same even when the details get more complicated.

Having indicated some of the details behind the probability calculations, I now want to return to the calculation of the mean for a Hardy golfer on a par 4. The key is to realize that some excellent shots are wasted. For example, the sequences ENN and ENE both represent birdies. The partial sequence EN creates the possibility of a wasted E, because a second E does not reduce the score at all.

These wasted Es raise the mean above 4 to $4 + u$, where u is the probability of a wasted E. To compute u, rewrite it in the form $u = pw$, where w is the probability of having a partial sequence like EN that creates the possibility of a wasted E. If the next shot is an E (probability p), then the E is wasted. The probability of being in a position to have a wasted E equals

$$w = 2p(1 - 2p) \sum_{k=2}^{\infty} \frac{k(k-1)}{2} p^{k-2}$$

$$+ (1 - 2p)^3 \sum_{k=3}^{\infty} \frac{k(k-1)(k-2)}{6} p^{k-3}.$$

The logic is that the partial sequence can have either one normal shot N (e.g., EN) or three normal shots (e.g., NNN).

The first sum describes all of the possibilities for having one N. For example, a final score of 3 can follow partial sequences of EN or NE, occurring with probability $2p(1 - 2p)$. A score of 4 can follow partial sequences of EPN, PNE, ... (6 possible orders), occurring with probability $2p(1 - 2p)3p$. A score of 5 can follow partial sequences of NPEP, PEPN, ... (12 possible orders), occurring with probability $2p(1 - 2p)6p^2$.

The second sum describes all of the possibilities for having three Ns. For example, a score of 4 can follow NNN, occurring with probability $(1 - 2p)^3$. A score of 5 can follow NNNP, NPNN, ... (4 possible orders), occurring with probability $(1 - 2p)^3 4p$.

Some calculus shows that the sums are $\frac{1}{(1-p)^3}$ and $\frac{1}{(1-p)^4}$, respectively. The resulting expression for w simplifies to

$$1 - \left(\frac{p}{1-p}\right)^4,$$

and therefore the mean for a par 4 is given by

$$4 + p\left[1 - \left(\frac{p}{1-p}\right)^4\right].$$

And that sums up our situation nicely!

FIVE

Handicap Systems and Other Hustles

I could beat you with a shovel, a baseball bat, and a rake.

—John Montague to Bing Crosby

When a single player asks to join your threesome on the first tee, you usually want to find out how good he or she is. You might ask for the player's handicap. A more direct approach is, "What do you usually shoot?" My brother George once asked this question and received the unusual answer of "94."

This guy was a pleasant playing companion but turned out not to be as good as advertised, failing to break 50 on the front nine. After a decent drive on the 15th hole, he picked up his ball, thanked everybody for an enjoyable round, and walked away. It turns out that he was not mad or late for an appointment. This was his unique strategy for coping with the frustrations of being a novice golfer: he always stopped after hitting his 94th shot. Instead of fretting about breaking 100 or having to record some astronomical number, he could enjoy his round and stop before it got ugly. His goal was to play the 18th hole.

I have wondered how a match against this fellow would work. I know I could give him a handicap (let him subtract strokes from his score on certain holes), but how many strokes would I give him? It might not help him much to receive strokes on the 17th hole.

One of the most ingenious handicaps ever offered was famously suggested by amateur John Montague to Bing Crosby, in the early 1930s, after a series of the crooner's "19th-hole" complaints about not receiving enough strokes. Bing was, at the time, a 3-handicapper, getting 5 or 6 shots but losing consistently to Montague. They headed back out to the 10th tee, Crosby with his golf clubs and Montague with a shovel, baseball bat, and rake retrieved from his car trunk. While Crosby was making a routine par, Montague tossed his ball into the air and hit it with the bat into a greenside bunker on the 366-yard hole. He then shoveled the ball on to the green, lay down with the rake, and used the handle like a pool cue to run the ball into the hole for birdie. Game over.

Undoubtedly, the Crosby family enjoyed a much prouder moment some 50 years later, when Bing's son Nathaniel won the 1981 U.S. Amateur.

Handicap systems of a more standard nature are the subject of this chapter.

A Wee Wager

Betting has been a part of golf since its beginning. A lively account of some of this history can be found in Michael Bohn's *Money Golf: 600 Years of Bettin' on Birdies*. One of the earliest references to golf is an edict from King James II of Scotland banning golf in 1457. The official reason for the ban was that the young lads of Scotland were being distracted from their archery practice (i.e., military training). Bob Cupp's book *The Edict* gives an entertaining alternative motivation for the edict, while describing what golf was like in its infancy.

The purpose of a handicap is to allow two players of different abilities to play and bet on equal footing. That is, each player should be equally likely to win their bet. For example, if I usually shoot 80

and you usually shoot 74, you might give me a 6-stroke handicap. Then, our average scores would tie, but a better-than-average 77 by me (with 6 strokes, a net $77 - 6 = 71$) would beat an average 74 by you.

The challenge is to determine how many strokes would be fair. With the exception of the Man Who Shoots 94, none of us have the same score every time out. Even if we did, a 94 from the white tees at the local municipal is not the same as a 94 from the back tees at Bethpage Black. So, a handicap system must take into account a variety of scores played on a variety of courses to produce a number that "fairly" represents a golfer's ability.

The USGA Handicap System

The handicap system used in the U.S. is administered by the United States Golf Association.[1] At first glance, it might seem impossibly complicated. One goal of this chapter is to explain each component of the system so that you can see its logic and its flaws.

Start with a score (S) for a round. The course that you played and the tee locations that you used have been assigned a course rating (CR) and a slope rating (SR) that indicate how difficult the course is. Essentially, the course rating tries to define the true par for the course. The slope rating is a correction factor that will be explained later.

The following computation gives you a *differential* (D) for your round.

$$D = (S - CR)\frac{113}{SR}$$

As an example, suppose that you shoot 82 on a course with rating 68 and slope 106. The differential for this round is

$$D = (82 - 68)\frac{113}{106} = 14.9,$$

using the convention that the differential is always rounded to the nearest tenth. For your USGA handicap, you take the differentials of your last 20 rounds and drop the 10 highest differentials. Your handicap is 96% of the average of the 10 smallest differentials.

The most striking feature of the USGA's system is the use of the best 10 rounds of the last 20. Different systems could use the middle 10 rounds (throw out the best 5 and the worst 5) or the worst 10 or even all 20 rounds. The choice of which scores to count has a large effect on the handicaps produced. The following example illustrates a flaw in the handicap system.

Suppose that golfers A and B post the following scores.

A: 71, 77, 71, 74, 74, 76, 79, 73, 73, 74, 74, 71, 74, 73, 74, 78, 73, 69, 77, 74

B: 80, 66, 79, 75, 78, 89, 68, 76, 72, 72, 80, 74, 72, 82, 78, 75, 74, 69, 77, 71

To keep things simple, suppose that for each round the course rating is par 71 and the slope rating is an average 113. Then the differentials are simply the actual scores with par 71 subtracted—in other words, the number of strokes above or below par.

A: $0, 6, 0, 3, 3, 5, 8, 2, 2, 3, 3, 0, 3, 2, 3, 7, 2, -2, 6, 3$
B: $9, -5, 8, 4, 7, 18, -3, 5, 1, 1, 9, 3, 1, 11, 7, 4, 3, -2, 6, 0$

The next step is to drop the 10 highest differentials for each golfer.

A: $0, x, 0, x, x, x, x, 2, 2, x, x, 0, x, 2, 3, x, 2, -2, x, 3$
B: $x, -5, x, x, x, x, -3, x, 1, 1, x, 3, 1, x, x, 4, 3, -2, x, 0$

Then average the remaining differentials and multiply by 0.96.

$$\text{A: } (0.96)\frac{0+0+2+2+0+2+3+2-2+3}{10} \approx 1.2$$

$$\text{B: } (0.96)\frac{-5-3+1+1+3+1+4+3-2+0}{10} \approx 0.3$$

Player A would have a handicap of 1, and player B would have a handicap of 0 (a "scratch" golfer).

There are some important observations to make before moving on. First, notice that player A receives a 1 handicap in spite of shooting 2 strokes or more over par in 16 out of 20 rounds. Second, player B is a scratch golfer in spite of recording 4 rounds in the 80s, including an 89! This explains the following description of the USGA handicap system: the USGA handicap measures **potential**, not average score.

You may have recognized the scores used in this example as being from the previous chapter, where player A is a fairly consistent player ($p = 0.1$), and player B is a more erratic player ($p = 0.2$). The erratic nature of B's play creates more potential for low scores, and this is reflected in the lower handicap. Player A matched or bettered his handicap of 1 only 4 times out of 20, while player B matched or bettered his handicap of 0 only 4 times out of 20. You can only expect to match or beat your USGA handicap 20 to 25% of the time.

Is It Fair?

If the purpose is to even out a match between two players, you can see that the USGA system does not work.[2] In the above example, player A beats player B 11 out of 20 times with 2 ties, yet player A would actually get a stroke from player B! With the stroke, player A now wins 13 out of 20 matches, with 1 tie. Player A's average score is one stroke better than player B's average score, but player A gets the handicap stroke. The handicap does the exact opposite of its objective: to even up the match.

You can object that the above numbers apply only to one special, made-up case. What can we expect for real golfers? The answer, as always, depends on the nature of the golfers. The lesson to learn from players A and B here is that the USGA handicap is strongly influenced by how erratic the golfer is. If a player has a 6 handicap, you should not expect that person to shoot 6 over par. If the player is unusually consistent, you might expect to see a score of 6 or so over the course rating (depending on the slope rating, of course). However, a more realistic expectation is that the player's "A" game will produce a score of about 6 over the course rating.[3] Golfers who do not bring their A game will score much higher than their handicaps would predict.

A couple of generic conclusions can be drawn. If two players have the same handicap, the more consistent player will have a lower score more often than not. It is then only a small step to this conclusion: for two players of different handicaps, the more consistent player will win more than half the time. Generally, a 10-handicap player will be more consistent than a 20-handicap player, since consistency is part of what makes a good golfer good. So, with appropriate "ifs" and "buts," we can say: in general, the USGA handicap favors the better player in head-to-head competition. One case, however, in which the "better" player is not favored is if the higher handicap player is more consistent.

Why would the USGA choose a handicap system like this? Part of the answer could be to discourage the reporting of bogus high scores. I have always wanted two handicaps: a low one to brag about and a high one to help win matches. A dishonest person could intentionally throw away shots on the course or shoot a bad round to try to artificially raise his handicap, but the USGA system requires such a person to cheat on the best 10 rounds of the year, which most golfers would not be willing to ruin.

More importantly, handicaps are not used just for head-to-head matches. They are also used for tournaments. A "captain's choice" ("scramble") tournament, where each group plays the best of four shots in the group, is more about potential than average. The group plays the best shot and picks up the bad shots.

Some tournaments are based on net score (actual score minus handicap). In this case, fair might mean that there is an equal chance of a low handicap player and a high handicap player winning. Recall from chapter 4 that, in a tournament of 90 player As and 50 player Bs, one of the inconsistent player Bs would pull off a career round and win 56% of the time. While most player As would have a better score than their player B counterparts, player As are generally too consistent to go low enough to win. In this case, it makes sense for player A to have a higher handicap as it makes up for the absence of player A's potential to match player B's best score.

So, what is the bottom line? The USGA handicap system does not work very well for evening up head-to-head matches, but it works well in large, stroke-play tournaments. The better, more consistent player has an advantage in head-to-head competition: keep this in mind when making bets.

Regardless of these biases in the system, the USGA works very hard to administer its system as fairly as possible. This is the reason that the system appears to be so complicated. In particular, the slope rating of courses is the result of a much-needed correction to the old system.

Different Courses

It is obvious that a good handicap system needs to rate the difficulty of a variety of courses. It is probably not so clear why there is both a course rating and a slope rating. You might be surprised to learn the reason: the USGA recognizes that professional golfers and average golfers have different needs.[4]

Imagine two courses, one Easy and one Hard. When pros play the two courses, the average scores are 70 on Hard and 64 on Easy. So, Hard should have a course rating that is 6 above Easy's rating. This makes sense, but wait. The drive on the third hole on Hard requires a 220-yard carry over water. This is no problem for the pros but would be a nearly automatic 2-stroke penalty for the average player. Hard has narrow fairways and tall rough, making it tough for the pros to reach some greens and costing them half a stroke on some holes when they drive into the rough. However, average golfers are doing well to reach the fairway chipping straight out from the rough. Each shot into the rough costs them at least a stroke, and they may chip out to the fairway from a wayward drive, only to have to chip out again when they flail a long iron right back into the rough.

The point is that hazards and features that increase scores a little for the professional can increase scores *a lot* for the average golfer. The purpose of the slope rating is to adjust the course rating for all levels of golfers. Compared to the Easy course, the Hard course may play 6 strokes higher for pros, but it might be 16 strokes harder for the average golfer and 26 strokes harder (almost unplayable) for a weak golfer. The handicap system tries to take this into account.

A Slippery Slope Rating

A thought experiment illustrates the logic of the slope rating system.[5] Imagine a course on which a scratch golfer averages 70 for the best 10 out of 20 rounds. Then 70 becomes the course rating. You might expect that a 5-handicapper would average 75 on this course. However, as noted above, the handicap of 5 counts only the best 10 out of 20 scores and is further reduced by a factor of 0.96. Taking this into account, it might turn out that 5-handicappers actually average 75.6, 10-handicappers average 81.3, 15-handicappers average 87, and 20-handicappers average 92.6.[6] We can visualize these results graphically, plotting the handicap on the horizontal axis and the average score on the vertical axis. When the points are plotted, they fall on a line with a slope of 1.13, as shown in figure 5.1.

The slope of 1.13 becomes the course's slope rating of 113—a rating considered "average" by the USGA. A harder course, with long carries over water and other punitive features, would have a different distribution, with some scores skyrocketing for high-handicap players. The assumption is that the scores would line up somewhat like those in figure 5.1. The slope rating for the course

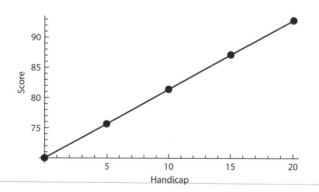

Figure 5.1 Hypothetical average scores for players of various handicaps on a course of average difficulty

is named for, and equals, the slope of the line that the scores form, multiplied by 100. In practice, a USGA rating team will establish the course rating and a rating for a bogey golfer (one who shoots 1 over par on most holes). These two data points determine a line, and the slope of that line determines the slope rating.

The hypothetical scores shown above are based on a general rule of thumb that the average of a golfer's top 10 out of 20 scores will equal about 92% of the average of all 20 scores. If the golfer's scores are normally distributed (that is, if they follow a bell curve), then the average of the best half of the scores should be about 0.8 standard deviations below the mean. For a golfer with an average score of 80 and a standard deviation of 8, the average of the best 10 out of 20 would be about $80 - 0.8 * 8 = 73.6$, which equals exactly 92% of 80.[7] Using the 92% rule, a golfer whose average score is A will have a handicap of $(0.92)(0.96)A = 0.8832A$. Conversely, if the player's handicap is H, then the average score would be $\frac{1}{0.8832}H \approx 1.13H$. The average slope rating of 113 therefore corresponds to a course on which the 92% rule holds.

As mentioned earlier, the differential that is actually used to compute your handicap is

$$(S - CR)\frac{113}{SR},$$

where S is the recorded score, CR is the course rating, and SR is the course's slope rating. As an example, suppose that you shoot 85 on a course with course rating 70 and slope 130. To compute your differential for this round, start with $85 - 70 = 15$, meaning that your score is 15 strokes above the course rating. The slope rating of 130 means that the course is more challenging than average. The expectation is that you would have had a better score playing on an easier course. The fraction $\frac{113}{130} \approx 0.87$ estimates how much better the score would have been. Since $15 * 0.87 \approx 13$, the prediction is that you would have scored 83 on a course of average difficulty, 2 strokes better than your actual 85. (Perhaps you did not make the 220-yard carry on the third hole.) Your differential for the round is 13, and it counts as one of the 20 differentials that determine your handicap.

The process is reversed when you play a match. That is, if you bring a 13 handicap to a course with a slope rating of 130, your effective handicap for the round will be $13 * \frac{130}{113} \approx 15$. Playing a harder course, as defined by the slope rating of 130, you need more strokes to compete on an even basis with better golfers.

Variants

There are numerous proposals for alternative handicap systems, and there are numerous handicapping systems in use around the world. The USGA system is used only in the United States and Mexico. For instance, the Australian Men's Handicapping System starts with an initial handicap determined by the results of the most recent 5 rounds, all of which are counted. This is done on the basis of differentials similar to those in the method described above, except that slope ratings are not used. From this starting value, adjustments are made to the handicap as more scores are recorded.[8] Although the rules are somewhat complicated, essentially a score that is higher than predicted by the handicap is used to increase the handicap by a small amount. In the same way, a score that is lower than predicted by the handicap is used to decrease the handicap by a small amount.

The USGA system has undergone quite a few changes in the past 100 years. The 96% factor used in computing handicaps was set at 85% until 1976.[9] The change to 96% was one of many compromises as the pooh-bahs of golf debated whether the handicap should reflect potential scores or average scores. The 96% figure splits the difference between the previous figure of 85% and that from a study commissioned by the USGA showing that it would take an adjustment of 107% to even out head-to-head matches.[10]

A commonly used handicap system is to self-report a handicap on the first tee and then adjust it at the turn if the match has become one-sided. The effectiveness of this system depends on what caused the imbalance in the first place. If one player had an unusually good nine holes, then the fairness of the adjustment depends on whether that person continues to play above average, returns to his or her normal game, or gets carried away by a temporary burst of competence and tries several risky plays that backfire.

The examples of Hardy golfers in the previous chapter show that more inconsistent golfers tend to have a higher scoring average but a greater likelihood of a very low score. This creates the apparent paradox of a player with a lower handicap having a higher scoring average. Recognizing the importance of variability in a golfer's scores, Bingham and Swartz have proposed a handicap system that takes both average and variability into account.[11]

The Back Tee: Variance

To see how variability affects results, consider two golfers of different ability, one averaging 80 and the other 90. Bingham and Swartz estimate that the scores of such golfers might have standard deviations of 3.27 and 3.8, respectively. Figure 5.2 shows graphs of the probability density functions (pdf's) for these golfers' scores.[12]

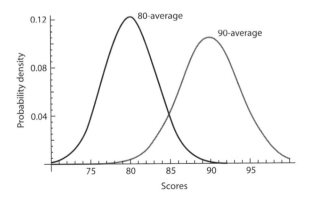

Figure 5.2 Graphs of probability density functions for scores of two golfers, one with mean 80 and standard deviation 3.27, the other with mean 90 and standard deviation 3.8

The bell-shaped curves assume that each player's scores follow a normal distribution. There is evidence that this is a reasonable assumption.[13] The numbers on the vertical axis are significant only in a relative sense. The higher the curve is for a given score, the larger the probability of the player achieving that score. The average scores of 80 and 90 locate the peaks of the curves. The standard deviation measures how "spread out" the curves are. Notice that the curve centered at 80 visibly separates from the axis only between 70 and 90, indicating that this golfer is very unlikely to score lower than 70 or higher than 90. The 90-average golfer has a little more variability, with likely scores extending down about 12 strokes to 78 and up 12 strokes to 102.[14]

Given the averages of 80 and 90, it might be that our two golfers have handicaps of 10 and 20. Subtracting these handicaps gives us the net scores of the two golfers (shown in figure 5.3). This graph shows more clearly that one distribution is more spread out than the other. The taller curve represents the low-handicap, consistent player who is not likely to stray very far from a net 70.

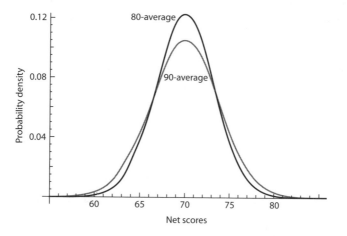

Figure 5.3 Probability density functions for net scores of the golfers in figure 5.2, assuming handicaps of 10 and 20

The high-handicap player is less consistent and has a wider range of possible scores.

The scenario that Bingham and Swartz explore is the wonderful day on which both players are playing well. The big question is how we should precisely define two players of different ability as both playing "well." This could mean 10 strokes better than average, but certainly an improvement from 80 to 70 is more meaningful than an improvement from 90 to 80.

Probability theory gives us a nice way to think about how to define "well." Let's say that playing well means having a score in the top 16% of all scores. In other words, only about 1 in 6 rounds is going to be this good. This hypothetical value is convenient because of the property of the normal distribution, whereby 68% of the scores are within one standard deviation of the average. For the low-handicap player, 68% of the scores are in the range 80 ± 3.27, placing the net scores between 66.73 and 73.27. For the high-handicap player, 68% of the scores are in the range 90 ± 3.8, placing the net scores between 66.2 and 73.8. For each player, about 16% of the scores will be less than the lower boundary. So, the 16% mark for the 10-handicapper is at 66.7, and the 16% mark for the 20-handicapper is at 66.2. The 20-handicapper is more likely to have a lower net score when playing well.

If the handicaps effectively equalize the scoring averages, then the less consistent player has a better chance of having the lower net score. The best 20% of net scores for the inconsistent player are better than the best 20% of scores for the consistent player. Of course, the exact same advantage flips to the consistent player if we look at the worst 20% of scores.

As we have seen, the USGA handicap is *not* designed to even out the scoring averages. A player averaging 90 might have an 18 handicap,[15] while a player averaging 80 might have a 9 handicap. Adjusting figure 5.2 by these amounts gives figure 5.4.

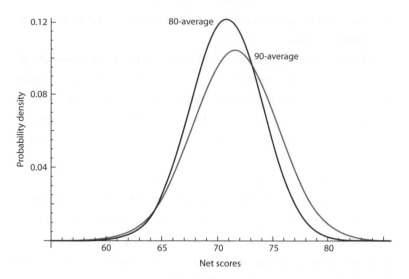

Figure 5.4 Probability density functions for net scores of the golfers in figure 5.2, assuming handicaps of 9 and 18

The advantage has now shifted primarily to the low-handicapper. The average net score is lower, and only for scores lower than 64 is there even a slight advantage for the high-handicapper. However, if many such players are competing in a tournament, this slight advantage translates into a good chance that one of the high-handicappers will post the lowest score. Bingham and Swartz computed the minimum net score out of 20 scores for a number of golfers at a club in British Columbia. They found that, on average, an increase in handicap of 10 reduced the minimum net score by one stroke.[16] This is strong evidence that, for players having their best round of the year, the USGA handicap system favors the high-handicapper.

The problem identified here is that for a tournament with a large number of players a handicap system should equalize players of different abilities who are playing their best rounds of the year. To know what the best round might be, you need a graph like figure 5.2 showing the range of possible scores and their likelihood. Then, you need a way of equalizing different players.

Statistically, there is a simple fix for this problem. In any introductory statistics class, you learn about standard scores (or z-scores). For a normally distributed random variable X, the average and the standard deviation uniquely determine the distribution. The z-score subtracts the average A and divides out the standard deviation SD. Then the quantity

$$z = \frac{x - A}{SD}$$

is normally distributed with average 0 and standard deviation 1. A comparison of z-values from different normal random variables is "fair" in the sense that a z-value of 1 always represents a score at the 16% mark, regardless of the original distribution.

In terms of golf handicaps, Bingham and Swartz equate x with the differentials computed above, A with the player's (unknown) true average, and SD with the player's (unknown) true standard deviation. They develop estimates for A and SD based on the player's USGA handicap H, ending with the formula

$$T = \frac{113(S - CR)/SR - 2.10 - 1.082H}{2.74 + 0.053H}.$$

Each player would plug in the score S, handicap H, course rating CR, and slope rating SR. The smaller the T-value, the better. Simulations run by Bingham and Swartz indicate that this calculation produces a match that is fair in the sense that there is no bias toward high- or low-handicap golfers for either a head-to-head match or a tournament.

And some people think that the USGA handicap system is complicated! For casual rounds, most golfers will probably continue to determine handicaps through negotiations on the first tee, but beware of anyone who just happens to have a baseball bat, shovel, and rake in his car.

PART TWO

Analysis of PGA Tour Statistics

The second part of this book focuses on the professional game. The PGA Tour's ShotLink system provides detailed (to-the-inch) information about the location of the ball before and after each shot, and a statistical analysis of this data informs us about what it takes to succeed professionally.

The data may be summarized by various statistics for the Tour. These statistics give information about which skills (putting, driving, and so on) are the keys to success and which skills are strengths or weaknesses for the various players. The individual skill ratings are combined into an overall rating system for professional golfers. Among the questions addressed:

- "Drive for show, putt for dough" implies that putting is more important than driving. Is that true?
- Which basic statistic is the best predictor of a final score?
- At what distances do the pros make fewer than half of their putts?
- Do the pros have more success putting for par or for birdie?
- Do the pros get the ball closer to the hole from the rough or from bunkers?

- How much accuracy is lost when hitting from the rough compared to the fairway?

- Does the average score on a hole depend on the length of the hole?

- Which players are the best at hitting different types of shots? Do the same ones dominate every year?

- How much better was Tiger Woods than everybody else in the 2000s?

To answer these questions, the data for each shot by each player are analyzed to compute how much better or worse than average the individual shot is. The resulting values for all shots of a given type (putt, chip, approach, and so on) are then added to obtain ratings for each player for that skill. The players can then be ranked on individual skills, and an overall rating is obtained by adding the skill ratings.

SIX

The ShotLink Revolution
Golf Statistics

Some of these numbers acquire a kind of poetry to them.
—*Tim Wiles, Baseball Hall of Fame*

Televised golf has changed in numerous ways, as technology has given us spectacularly detailed slow-motion replays, swing analyses, and shot tracers that show the curve of a ball in flight. The level of precision provided is also revolutionary, if not as visually dramatic. Instead of an on-course announcer guessing that a putt is "at least 30 to 35 feet," the announcer in the booth can authoritatively say that the putt is 33 feet, 5 inches. The source of this precision is ShotLink, and it is rapidly transforming the world of golf statistics.

I should hasten to emphasize that this is statistics with a lowercase "s." My dad was a Ph.D. statistician and practiced Statistics with an upper-case "S." Research statisticians employ sophisticated, highly mathematical techniques to build accurate models from carefully collected experimental data. In the early 1960s, my dad was also a statistician for the Dallas Cowboys. This meant that he carefully accounted for each play in a game and did the arithmetic to compute Don Meredith's passing percentage. The golfing equivalent of such statistics is discussed in the next several chapters.

Baseball's Poetry

Baseball fans have a special affinity for the statistics of their game. Certain numbers are so recognizable that the Baseball Hall of Fame will not correct players' plaques when new information creates a change in the record books. The Hall of Fame does not want to disrupt the "poetry" of the museum, and indeed baseball stats could make a good *Jeopardy* topic, perhaps called "This Year in Baseball." Try your luck with these (answers are in the notes):

The year of .406 and 56.
The year of 70, 66, and 56.

The year of 1.12, 31, $58\frac{2}{3}$, and .301.

The year of 61 and 54.[1]

The popularity of baseball cards owes a large debt to the statistics on the back of each card,[2] stats that generations of fans learned and memorized. The main reason why baseball has such great number recognition is the discrete nature of the game. This is different from "discreet," which baseball players have only rarely been but reporters used to be. In this setting, discrete means that each pitch, at bat, and so on is a separate event that can be listed and quantified. Compare this to European football or interior line play in American football, where the action is fluid and continuous and, therefore, difficult to quantify. Golf is also discrete, since a round consists of 70 or so separate strokes—for professionals, anyway. However, baseball plays always start with the ball in the pitcher's hand and the batter at home plate, while a golf stroke can be taken from almost anywhere on the course and (sadly) can end almost anywhere on or off the course. The limitless options for ball location make golf much more difficult to describe numerically.

Baseball statistics were revolutionized by Bill James. The annual *Bill James Baseball Abstracts*, published in the 1980s, introduced a new way of looking at baseball. James found improvements to classic statistics like batting average and RBI (runs batted in), quantified the number of runs an individual player contributed offensively, and invented ways to measure a player's defensive contribution to a team. Frustrated with his inability to obtain detailed baseball data, James organized Project Scoresheet, which collected data on every pitch and batted ball of each new season. This data opened the floodgates as thousands of fans could now work out their own special "sabermetrics"[3] theories.

By contrast, the golf revolution came from above. There was no Bill James rattling the saber (so to speak) for better data. The world of golf statistics was completely revolutionized by ShotLink (see logo, figure 6.1), the PGA Tour's system of lasers and volunteers who record data. ShotLink determines the starting and finishing locations of each shot *to the inch*. We can know that Davis Love III's second shot on the first hole of the 2007 Mercedes-Benz Championship started in the left fairway 8,156 inches from the hole and finished in the right fairway 1,090 inches from the hole.

Before we dive into the data, there is an important disclaimer to be made. ShotLink is administered by the PGA Tour,[4] which runs many of the professional tournaments in the United States but not all. The statistics that follow come from ShotLink only, so some tournaments are not included. The most important absences are the four men's major golf championships[5] and tournaments from the LPGA, European, Japanese, and other tours.

Those Good Old-time Statistics

If you track your own golf statistics, you probably record your number of putts, fairways hit, greens hit in regulation, and up-and-downs for each round. These basic statistics are relatively easy to mark on a scorecard. Similar statistics have been available for PGA golfers for years.

Greens hit in regulation is clearly a useful measure of solid ball-striking. In 2009, ShotLink recorded data for 307,836 holes. On 199,161 of these, the golfer hit the green in regulation (that is, after a standard number of strokes). This computes to a 64.7% rate for hitting greens, a percentage that has changed very little since

Figure 6.1 ShotLink, a revolutionary data collection system

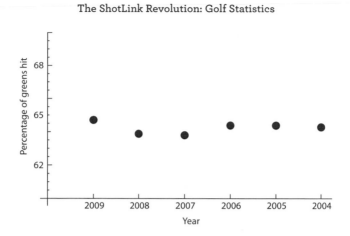

Figure 6.2 Percentage of greens hit in regulation, 2004–2009

2004 (figure 6.2). Even with the magnified vertical scale of this graph (showing only 60 to 70), it is clear that the percentage has not deviated much from 64%. While the overall Tour average has stayed nearly constant, percentages for individual golfers can increase or decrease significantly from year to year. Table 6.1 shows the top five percentages of greens hit in regulation during 2004 through 2009. Some players repeat on the list, but others rank near the top one year and then drop back into the pack.

The consistency of a statistic will be a concern for us. When a statistic varies wildly from one year to the next, we question whether that statistic measures a skill level or is simply random noise.

Nuggets and Flakes

Bill James is said to have written, "Do we need to have 280 brands of breakfast cereal? No, probably not. But we have them for a reason—because some people like them. It's the same with baseball statistics."[6] With ShotLink, we have thousands of statistics available: percentage of putts made from 12 feet, average approach distance for shots from the fairway 100–110 yards from the hole, and so on. Which of these statistics provide important information? One goal

Table 6.1 Top fives in greens in regulation, 2005–2009

Year	Player	Greens hit in regulation (%)
2005	Garcia	71.8
	Durant	70.9
	Perry	70.8
	Singh	70.5
	Brehaut	70.0
Tour average		64.4
2006	Woods	74.1
	Gove	72.0
	Senden	71.1
	Furyk	70.7
	Baird	70.2
Tour average		64.4
2007	Woods	71.0
	Senden	70.3
	Gove	70.2
	May	70.1
	Beckman	70.0
Tour average		63.8
2008	Woods	71.4
	Byrum	71.1
	Durant	71.1
	Allenby	70.4
	Frazar	70.3
Tour average		63.9
2009	Fisher	71.7
	Stanley	71.5
	Senden	70.9
	Byrd	70.6
	Durant	70.6
Tour average		64.7

of this chapter is to separate the statistics with good nutritional value from the ones that are just empty calories.

The ultimate goal in a round of golf is to post the lowest score possible, so a criterion for a useful statistic is that it must relate to scoring. As a first attempt at evaluation of the available statistics, I computed the correlation between several statistics and scoring. For this study, the data were sorted by tournament. Then, for example, the list of numbers of putts taken for the tournaments was correlated to the list of scores for the 6,000-plus player tournaments in the data set for a given year. In 2009, the correlation rounds to 0.238. I then created a list of number of greens hit in regulation for each tournament. The correlation between greens hit in regulation and score in 2009 is −0.579. A scatter plot of the data for greens in regulation is shown in figure 6.3.

The correlation between two variables gives information about how the variables change from data point to data point. The sign

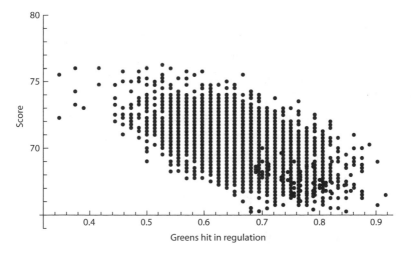

Figure 6.3 Scatter plot of average tournament score versus greens hit in regulation, 2009

(+ or −) of the correlation relates to whether the values of the variables increase or decrease. The negative sign for greens hit in regulation and score means that, in general, when the number of greens hit goes up, the score goes down. This certainly makes sense and is clearly visible in figure 6.3. The positive sign for putts and score means that, when the number of putts increases, the score increases. This is also what we would expect. With equal validity, we can say that the correlation of −0.579 means that, when the score goes up, the number of greens hit goes down. That is, correlation simply relates the two variables, in either order. Correlation does not imply causation. In this case, it makes sense that hitting more greens can cause your score to drop and that taking more putts clearly causes your score to increase. However, it is very important to realize that a high correlation in no way *proves* that changes in one variable cause changes in the other.[7]

The numerical value of the correlation is also meaningful, although the meaning can be subtle. A correlation is always between −1 and 1. A correlation of 0 indicates that there is no measurable linear trend in how the quantities increase or decrease. Figure 6.4a shows a plot of data with the coordinates having a correlation of 0.09. An increase in one variable is equally accompanied by increases and decreases in the other variable. Correlating with score, a correlation close to 0 indicates a statistic that does not predict scoring well (using a linear equation). A high correlation (close to 1 or close to −1) indicates a statistic that could be used to accurately predict scores and, in this sense, is an important statistic. Figure 6.4b illustrates a correlation of 0.36. This is not very close to 1, but you can see a general trend for the points to be higher (increased y) as you move from left to right (increased x). You might imagine a line through the middle of the data points which could be used to predict scores.

To get a little more technical, correlation measures the extent to which there is a linear (straight line) relationship between the

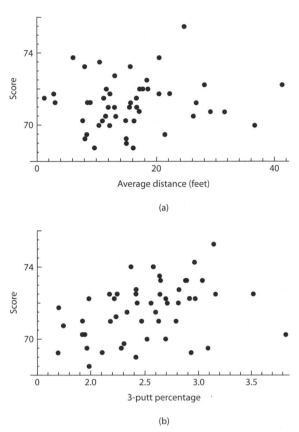

Figure 6.4 Low versus higher correlations: (a) low correlation (.09); (b) higher correlation (.36)

variables. The points in figure 6.4b come closer to forming a line than do the points in figure 6.4a. The square of the correlation gives the percentage of variation in the data that can be explained by the best-fit line. In figure 6.4b, there is a line (through the "middle" of the data) that explains $0.36^2 \approx 0.13$, or about 13% of the variation in the data. The best-fit line in figure 6.4a explains a mere $0.09^2 = .0081$, or less than 1% of the variation in the data. The best-fit line in figure 6.3 explains $0.579^2 \approx 0.335$, or about 33.5% of the variation. You can see that, as you move to the right (more greens hit in regulation), there is a tendency for the average score to drop, but there is a fairly wide band of scores at each value of greens hit in regulation. That is, a significant amount of variation in scoring (about 66.5%) is not explained by greens hit in regulation.

For our purposes, we will use the guideline that bigger correlations (ignoring the plus or minus sign) are better. I computed six-year (2004–2009) averages of correlations for a large number of statistics. Ranked by size of correlation, the top six statistics are given in table 6.2. Here, "scrambling" reflects only those holes on which the golfer misses the green and equals the percentage of times the player makes par or better.

The first three statistics are completely reasonable. To score well, you want to hit many greens and take few putts. If you miss the green, you want to get up-and-down to save par. Greens hit in regulation and putts per green hit in regulation (often called "putting average") are among the few statistics that are readily available online. Week by week, they are the two statistics that best predict overall score.[8] Notice that they are the only statistics with correlations above 0.5.

You might be surprised that the number of putts per green hit in regulation correlates to scoring better than the total number of putts does. The problem with the total number of putts is that it is a "combination" statistic that has multiple influences working

Table 6.2 Correlations of statistics to scoring, 2004–2009

Rank	Statistic	Correlation
1.	Greens in regulation	−0.591
2.	Putts per green in regulation	0.530
3.	Scrambling	−0.489
4.	3-putt percentage	0.301
5.	Percentage made of putts under 10 ft.	−0.290
6.	Proximity of approach shots	0.266

at cross purposes. While it is never good to take a large number of putts, one way to minimize the number of putts taken is to miss every green and chip close. It is easy to improve your total putts statistic without improving your overall scoring.

One type of statistic that is conspicuously missing from table 6.2 is a driving statistic. Modern technology allows players to adopt a "bomb and gouge" approach. Even if the drive misses the fairway, they have bombed it close enough to the green that they can gouge a wedge out of the rough and onto the green. Four common driving statistics have the following correlations (2004–2009) with scoring: driving distance (-0.193), fairways hit (-0.184), average distance (-0.151), and longest drive (-0.101).

Driving distance has a slightly higher correlation than percentage of fairways hit, but the correlation to scoring is not very high for either.[9] There are actually two driving distance statistics. Distance is measured on all holes, but values are separated out for holes on which most players hit driver. Driving distance is for the special driving holes. The correlation to scoring for tee shot distance on all holes (average distance) is slightly less.

Driving distance, then, does not seem to have much to do with scoring well.[10] The bottom line seems to be that, on the PGA Tour, driving is important only insofar as it increases or decreases your ability to hit the green in regulation. A drive that leaves you stuck behind a tree obviously affects your score adversely. However, on the larger scale of multiple rounds, the evidence does not support the theory that driving is a critical factor in scoring well.

The sixth best statistic in table 6.2 is "proximity of approach shots," which is the average distance to the hole after an approach shot. This statistic would clearly influence each of the top four statistics, since the closer you are to the hole the more likely you are to be on the green and have a chance to make a birdie putt. I find it interesting that proximity of approach shots has a much smaller correlation to scoring than does greens hit in regulation.

The proximity of approach shots can be broken down by distance. Aggregated into 25-yard intervals, the distance range with the highest correlation to scoring is 150–175 yards (a correlation of 0.267 for 2004–2009), followed by 125–150 yards (0.244) and 175–200 yards (0.217). The distances shorter than 125 yards and longer than 200 yards all had lower correlations, as shown in figure 6.5. One confounding factor for the shorter distances is the occasional need to hit these shots after laying up from a bad drive. That is, a good drive followed by a poor shot from 60 yards could result in a par, the same as a great shot from 60 yards after a horrible drive and pitch out. The lower correlations do not necessarily mean that the skills are less important, only that a smaller average approach distance for a tournament does not always correspond to a lower score for the tournament.

"Scrambling" ranks third on the list of statistics in table 6.2. Scrambling is defined as the percentage of times the golfer made par or better on holes on which the golfer did not hit the green in regulation. The tour average from 2004–2008 was 56.5%. Percentages

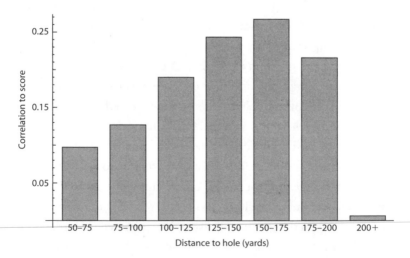

Figure 6.5 Correlations of approach distances to score

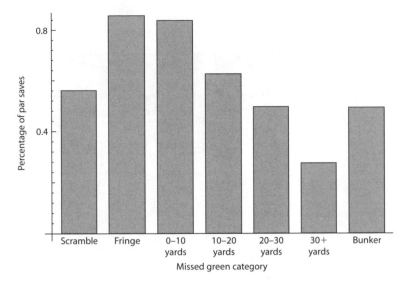

Figure 6.6 Percentages of par saves

of par saves from different categories are illustrated in figure 6.6. Notice that sand saves are about the same as saves from 20–30 yards (60–90 feet). Which of these scrambling categories is the most important? The category with the highest correlation to score is saving par from 10–20 yards, followed by sand saves. However, the correlations are all fairly small (see figure 6.7).

Three of the top six statistics (ranked by correlation with score) are putting statistics. This is one of several indicators that putting is the most important skill in golf. Putting statistics are the focus of the next chapter, however. For now, I want to complete the study of correlations by reporting scores for several putting statistics. Figure 6.8 shows correlations for six common putting statistics. To make comparisons more visual, the absolute values of the correlations are shown. Here is a conclusion worth repeating: Putts per green hit in regulation is the best basic putting statistic. Percentages of 1-putts, 3-putts, and putts per round have significantly lower correlations with scoring. Distance is the total length of all

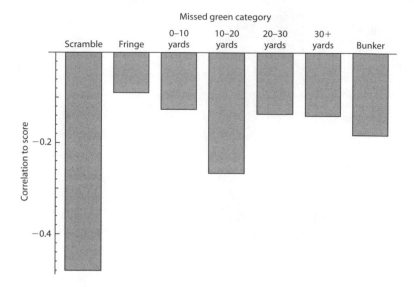

Figure 6.7 Correlations of par saves to score

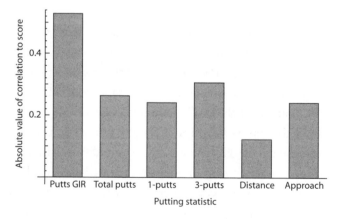

Figure 6.8 Absolute value of correlations of putting statistics with score

made putts. Approach is the average distance to the hole after first putts that did not go in. As you would expect, the correlations for 1-putts and distance are negative, while the other correlations are positive.

Some golfers are solid on short putts but cannot make long putts. Others make more than their share of long putts but are shaky on short putts. What can be said about which distance is the most important? The answer is "not much" using correlations. Broken down using the PGA Tour's distance ranges, the highest correlation is for putts from less than 10 feet. Direct comparisons are not entirely valid, since there are vastly different numbers of putts taken from the different distance ranges. Some of the dangers of small sample sizes are discussed next.

When Correlations Don't Relate

The following discussion is intended as a caution against over-using correlations. The story begins, unfortunately, with an overuse of correlations. The first time I ran the correlations for figure 6.9, the values for the ranges $15'-20'$, $20'-25'$ and $25'+$ were 0.111, 0.196, and 0.158, respectively. My first thought was that they were too large, and then I noticed that all three correlations were positive. In other words, the higher the percentage of long putts made, the *higher* the score!

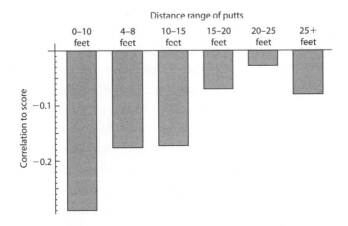

Figure 6.9 Correlations of percent of putts made from different distance ranges to score

Some detective work was required to figure it out. Here's what happened: I was using *all* tournament lines, whether the golfer made the cut or not. Some golfers who missed the cut happened to make the only putt they took from one of these distance ranges. If they had played more rounds or hit more greens in the rounds they played, they would have faced more putts from that distance and would have missed some. As it was, there were several 100% entries in the data, almost all associated with golfers who missed the cut (and, obviously, had high scores). There were enough such entries to shift the balance of the relationship to a positive correlation.

The values shown in figure 6.9 were compiled from tournament data using only those golfers who made the cut, which may affect interpretations of the correlations. Certainly, the correlations cannot be used to support any theories about which aspects of the game are most associated with missing the cut.

Given the above discussion, you may well question the value of −0.08 in figure 6.9 for putts of length greater than 25 feet. The correlations for 10′–15′, 15′–20′, and 20′–25′ show a nice trend whereby the longer the putt, the smaller the correlation with scoring. The argument that this trend should continue is based on my assumption that holing a long putt is a rare event, most often accidental. You do not have to make long putts to get a good score, and you can have a bad score in a round in which you make a long putt.

By this logic, the correlation for distances greater than 25 feet should be extremely small. However, it is not. A possible explanation is related to the argument for excluding golfers who miss the cut. The number of putts attempted from long distance is small enough that a lucky made putt could produce an unnaturally large percentage and distort the correlation. Another possibility is that the correlation (which is not large) is giving us some causal information. If you make a 30-foot putt, you have lowered your score one stroke below any reasonable expectation of what you should make on that hole. Holing a 60-footer not only beats a 2-putt by one; it also beats a likely 3-putt by two. That is, there is a very real savings when you make a long putt.

The primary lesson here is to be wary of placing too much importance on a single correlation between two statistics, especially if the number of data points is small.

The Back Tee: Tailoring the Basic Statistics

One objective of the second half of this book is to rate the PGA golfers. The work in this chapter sets the stage for a logical rating system. The bad news is that it is not a great rating system. The good news is that we will be able to improve on it in the chapters to come.

The correlations computed above identify three statistics as being especially good predictors of scoring: greens hit in regulation, putts made per green in regulation, and scrambling. These statistics have the advantage of being readily available online. The three variables can be combined in a linear best-fit model (also called *linear regression*) to give a better predictor of score. If GIR equals the fraction of greens hit in regulation, $PUTT$ equals the average number of putts per green hit in regulation, and SCR equals the fraction of par saves per missed green, then

$$S = 65.98 - 13.00\, GIR + 10.29\, PUTT - 7.94\, SCR$$

gives a good predictor of score. In turn, S (predicted score) can also be used to rate golfers. Plug in a golfer's stats, and the predicted score indicates the golfer's level of performance. The lower the predicted score, the better the golfer is.

The derivation of the formula for S starts with a general equation for a linear combination of the variables GIR, $PUTT$, and SCR. That means we want an equation of the form $S = a + b * GIR + c * PUTT + d * SCR$ for (unknown) numbers a, b, c, and d. Our task is to figure out the "best" values for the parameters a, b, c, and d. "Best" is in quotes because there are numerous ways to define precise criteria for optimality. The *least squares* criterion is commonly used.

To define the least squares criterion, start by imagining a golfer who hits 60% of the greens, averages 1.8 putts per green in regulation, and saves par 70% of the time. The predicted score for the golfer is $S = a + .6b + 1.8c + .7d$. If the golfer actually averages a score of 70, then the error in the prediction is $|a + .6b + 1.8c + .7d - 70|$, the difference between the predicted score and the actual score. Square the error and add up the squares of the errors for all of the golfers in the data set. The values of a, b, c, and d identified by the least squares criterion are the ones that make the sum of the squares of the errors as small as possible.

The description may sound complicated, but the mathematics for solving the least squares problem is surprisingly straightforward.[11] Spreadsheets and graphing calculators can do this easily. In the case of PGA tournament data for the 2004 through 2009 seasons, the values that minimize the total (squared) error turn out to be $a = 65.98$, $b = -13.00$, $c = 10.29$, and $d = -7.94$. The correlation between the predicted scores and the actual scores is slightly over 0.9, which is far higher than the correlation for any individual statistic.

So, what is wrong with this system? Both its strength and its weakness derive from the fact that it uses simple statistics. There are better statistics, which we will develop in later chapters. For

Table 6.3 Top five ratings from regression, 2009

Rank	Player	Predicted score
1.	Tiger Woods	69.6
2.	Steve Stricker	69.8
3.	Kevin Na	70.2
4.	David Toms	70.3
5.	Zach Johnson	70.3

now, let's look at how the system rates the PGA Tour golfers in 2009, using statistics from the regular season. For example, to get Tiger's predicted score, you plug in $GIR = 0.6849$, $PUTT = 1.738$, and $SCR = 0.6809$ and get $65.98 - 13(.6849) + 10.29(1.738) - 7.94(.6809) = 69.554$, which rounds to 69.6. The top five ratings for the 2009 season are shown in table 6.3.

There is not much arguing to be done with the top two; Tiger Woods and Steve Stricker finished one-two in the regular season FedEx Cup standings (a point system that the PGA Tour uses to rank Tour golfers). Zach Johnson also ranked in the top five in FedEx Cup points. However, Kevin Na was 18th and David Toms 19th in FedEx Cup points, and no experts ranked them in the top ten, much less the top five. The ratings in table 6.3 are therefore suspect.

Subjectively, then, this rating system is not bad, but neither is it great. It is not going to replace any of the official rating systems. The advantage of this system is that it uses readily available statistics. You can go online, grab some numbers, and plug them into the system at any point in the season. For this reason, it might be a useful system for choosing fantasy golfers.

Mathematically, we can say that this system is the best linear combination (in the least squares sense) of the three basic statistics that have the highest correlation with scoring. Translating that into everyday, imprecise, slightly exaggerated language: it is the best that we can do with basic statistics.

SEVEN

Lags and Gags
Putting Statistics

Unbelievable. I knew he'd make it.

—*Rocco Mediate, after Tiger Woods made a 12-foot putt
to tie him in the 2008 U.S. Open*

The 2008 U.S. Open at Torrey Pines was one of the most dramatic tournaments ever. Tiger Woods, hampered by a torn anterior cruciate ligament and stress fractures of his tibia, made a 12-foot birdie putt on the 72nd hole to tie Rocco Mediate and force a playoff. In the playoff the next day, Tiger again birdied the 18th hole to tie Mediate and finally won in sudden death on the 91st hole. The video of Tiger's tying putt on the 72nd hole is one of the most-played golf videos online, featuring replays from several angles and a wild celebration by Tiger and caddie Steve Williams.

One of the replays shows a close-up of the ball rolling and bouncing on its way to the hole, then catching the right edge and (barely) curling into the hole. This replay is a great illustration of the randomness involved in putting, which we first discussed in chapter 3. My favorite quotation about this putt is Rocco's comment in the epigraph. "Unbelievable" and "I knew he'd make it" do not logically go together, but somehow they capture many aspects of the scene. A normal golfer would not make that putt, yet everybody expected Tiger to make it.

There were conflicting aspects about that putt for Rocco. If Tiger had missed, Rocco Mediate would have become a U.S. Open champion, yet the playoff gave him another day in the national

spotlight, going one-on-one on national television with the best player of our time and matching him stroke for stroke. To his great credit, Rocco has consistently emphasized the thrill of the playoff rather than regret over his near miss.[1]

Thinking of all the dramatic putts that Tiger Woods has made over the years, the obvious question is whether he is the best putter on the Tour. We'll get part of the answer in this chapter and a more complete one in chapter 11. Compiling ShotLink data can show us that, in 2007, Tiger made 12 out of 27 (44%) of his 7- to 8-foot putts. But here is the big question. How good is 44% from 8 feet? Is it better than most PGA players typically do, right at the Tour average, or well below average?

These and related questions are the focus of this chapter.

Puttering Around

Looking at all putts on the 2007 PGA Tour (not including the four majors), it turns out that 165,026 putts were attempted from distances of 3 feet or less. A full 163,676 of these putts were made. This makes the Tour average of putts made from 3 feet or less

$\frac{163,676}{165,026} \approx 0.992$, or about 99%. From distances greater than 3 feet but less than or equal to 4 feet, the Tour made 30,653 out of 33,532 putts, or about 91%. Other percentages are given in table 7.1. Note that Tiger's 44% from 8 feet is well below the tour average.

The Tour percentages in other years are similar. One way to summarize the data in table 7.1 is to break it into quartiles of percentages. We see that:

- At all distances beyond 5 feet, the percentage of putts made is less than 75%.
- At all distances beyond 8 feet, the percentage of putts made is less than 50%.
- At all distances beyond 14 feet, the percentage of putts made is less than 25%.

A scatter plot showing distance in feet on the horizontal axis and percentage of putts made on the vertical axis is given in figure 7.1. Up to about 20 feet, the data points line up nicely on a curve. The curve $y = 148e^{-.122x}$ matches the data fairly closely and is superimposed on the data points.[2]

Table 7.1 Percentage of putts made from different distances, 2007

Length (ft)	Putts made (%)	Length (ft)	Putts made (%)
0–3	99.2	12	32.1
4	91.4	13	28.4
5	80.9	14	26.5
6	70.3	15	24.5
7	60.9	16	22.0
8	53.1	17	19.9
9	46.9	18	19.1
10	41.8	19	17.3
11	37.0	20	15.6

Source: Data from "event level" files from the PGA Tour.

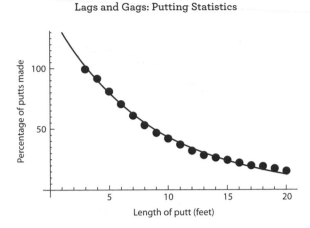

Figure 7.1 Percentages of putts made from 3–20 feet in ShotLink tournaments in 2007

A different way of looking at the putting statistics is to record the distance of the golfer's first putt and then compute the number of putts needed on the hole. For example, there were 12,833 times in 2007 that a PGA Tour player's first putt was between 4 feet and 5 feet long. In these 12,833 instances players took a total of 15,614 putts. The average number of putts starting between 4 and 5 feet away is then $\frac{15,614}{12,833} \approx 1.22$ putts. Figure 7.2 shows the scatter plot of data points for average number of putts from different distances, along with the curve $y = 0.88 + 0.337 \ln(x - 2)$. There is a milestone reached at 32 feet. Specifically, at all distances beyond 32 feet, the average is more than 2 putts.

Figure 7.3 shows the average number of putts for distances out to 100 feet, along with the curve $y = 0.88 + 0.337 \ln(x - 2)$. The figure illustrates a principle that we need to keep in mind in any discussion of statistical information. Starting at about 60 feet, the data points no longer follow a nice curve. The points seem to bounce above and below a curve running through the center. The reason for the random ups and downs of the data points is, in fact, randomness. Whereas there were 5,453 putts taken from 20 feet,

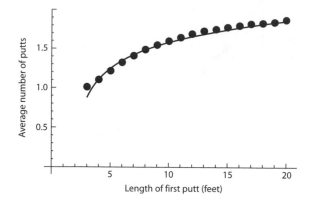

Figure 7.2 Average number of putts needed from 3–20 feet in ShotLink tournaments in 2007

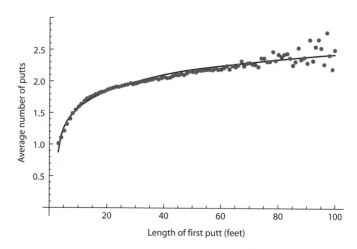

Figure 7.3 Average number of putts needed from 3–100 feet in ShotLink tournaments in 2007

there were only 8 taken from 100 feet. A lucky 1-putt or an unlucky 3-putt from 20 feet will not affect the average noticeably, but if one of the 8 putts from 100 feet goes in, the average can change dramatically. The sample size is critical.

Short Distance Gags

Having looked at overall Tour averages, we now want to see which golfers make the most putts. The ShotLink data allow us to break the putting distances down to the inch. We will start with a broader range, looking at all putts of less than 10 feet. This, you may recall from chapter 6, was the distance range with the highest correlation to scoring. Table 7.2 shows the top 10 putters on the Tour from less than 10 feet, for each of the years from 2007 through 2009.

What can we learn from these lists? Amazingly, only one player, Stewart Cink, made the top ten in more than one year, finishing 8th in 2007 and 9th in 2009. This is an indication that the percentage of putts made from less than 10 feet is not a stable statistic. One of the studies that I would like to do (or see done by others) is to analyze more thoroughly how putting statistics vary over time. During a season, are some putters streaky, with four or five good tournaments followed by several bad ones? Are the ups and downs in putting efficiency predictable or apparently random?

Table 7.2 Top ten percentages of putts made from 0–10 feet, 2007–2009

Rank	2007	Putts made (%)	2008	Putts made (%)	2009	Putts made (%)
1.	J. Parnevik	89.7	T. Purdy	91.4	P. Claxton	90.4
2.	K. J. Choi	89.5	S. Gump	89.5	J. L. Lewis	90.2
3.	Z. Johnson	89.4	J. Durant	89.5	T. Woods	90.2
4.	A. Oberholser	89.4	R. S. Johnson	89.4	J. Coceres	90.1
5.	B. Gay	89.4	B. Langer	89.3	K. Triplett	89.6
6.	G. McNeill	89.2	H. Mahan	89.3	B. Bryant	89.5
7.	C. Wi	89.2	S. Lowery	89.3	J. Byrd	89.5
8.	S. Cink	89.2	R. Garrigus	89.1	C. Pavin	89.5
9.	D. LaBelle	89.2	F. Jacobson	88.9	S. Cink	89.3
10.	A. Baddeley	89.1	R. Whitaker	88.9	J. Ogilvie	89.3

Table 7.3 Top ten percentages of putts made from short distances, 2004–2009

Rank	0–3 ft	Putts made (%)	3–4 ft	Putts made (%)	4–5 ft	Putts made (%)
1.	E. Els	99.8	C. Bowden	96.6	S. Stricker	88.9
2.	J. Cook	99.7	M. O'Meara	95.8	B. Faxon	88.4
3.	B. Faxon	99.7	J. Cook	95.8	B. Crane	88.1
4.	L. Donald	99.7	T. Woods	95.7	S. Gutschewski	87.5
5.	T. Woods	99.7	G. Willis	95.5	A. Oberholser	87.3
6.	P. Goydos	99.7	Z. Johnson	95.3	T. Woods	87.3
7.	C. Campbell	99.6	I. Poulter	95.3	I. Poulter	87.2
8.	J. L. Lewis	99.6	L. Mattiace	95.2	D. Wilson	87.1
9.	J. Driscoll	99.6	J. Parnevik	95.2	G. Coles	87.0
10.	P. Sheehan	99.6	F. Funk	95.2	J. Cook	86.9
	Average	99.2	Average	91.4	Average	80.9

Another way to parse the short putt data is to compile averages of made putts from smaller, 1-foot distance ranges. As you might expect, the results are even more variable than for the range of 0–10 feet shown in table 7.2. To try to get some meaningful information, I have averaged over multiple years to smooth out some of the year-to-year variability. Results are shown in table 7.3.

The fact that Tiger Woods and John Cook made all three lists means that they are among the best short putters in the game. Interestingly, Tiger does not necessarily jump out as a great putter if you look at year-by-year statistics. For example, from 2004 to 2008 his yearly rankings from 0–3 feet were 62, 1, 12, 80, and 153, respectively. From 3–4 feet, the rankings were 17, 57, 2, 56, and 160, respectively. As far as rankings go, then, an individual player's performance can vary substantially from one year to the next.

Middle Distance Sags

We next look at six-year averages from longer distances. One important issue here is the extent to which putting is a single skill. That

Table 7.4 Top ten percentages of putts made from medium distances, 2004–2009

Rank	10–15 ft	%	15–20 ft	%	20–25 ft	%
1.	L. Mattiace	39.1	B. Heintz	24.9	B. Faxon	18.8
2.	I. Poulter	36.6	W. Short	24.2	J. Parnevik	17.7
3.	B. Crane	36.1	T. Woods	23.7	T. Woods	17.7
4.	B. Geiberger	35.6	C. Riley	23.6	B. Geiberger	17.0
5.	L. Mize	35.3	D. Chopra	23.4	N. Green	16.7
6.	J. Delsing	35.2	S. Stricker	23.3	B. Crane	16.6
7.	A. Atwal	35.2	A. Atwal	22.6	S. Appleby	16.6
8.	M. O'Meara	35.1	D. Wilson	22.5	C. Pavin	16.4
9.	B. Faxon	35.1	L. Mattiace	22.5	A. Baddeley	16.4
10.	D. Wilson	34.7	J. Byrd	22.4	J.-M. Olazabal	16.1
	Average	31.0	Average	18.9	Average	12.7

is, do players who putt well from short distances automatically putt well from longer distances? Some golfers have been quite successful at short range with putting strokes that are essentially jabs, but this short, choppy stroke might not work well from longer distances. In addition, the ability to read breaks is more crucial from longer distances. From 3 feet away, you can often hit a putt hard enough that it does not break significantly. That is not typically an option from 15 feet away.

The top ten putters for distances longer than 10 feet over the combined years of 2004 through 2009 are shown in table 7.4. The percentages are amazing. Over a six-year span, Len Mattiace made nearly 40% of his putts from 10 to 15 feet. Bob Heintz made one-fourth of his putts from 15 to 20 feet. Tiger Woods, Brad Faxon, Brent Geiberger, and Ben Crane make the top ten on two of these lists.

Long Distance Lags

Over longer distances, the number of putts attempted by any given player decreases to the point where yearly percentages mean little.

Table 7.5 Top ten percentages of putts made from long distances, 2004–2008

Rank	25–30 ft	Putts made (%)	30+ ft	Putts made (%)	50+ ft	Putts made (%)
1.	S. Gutschewski	14.1	B. Pappas	6.75	W. Short	6.00
2.	G. Coles	13.2	F. Jacobson	6.37	T. Woods	5.34
3.	G. Willis	13.2	M. Gogel	6.27	B. Langer	5.19
4.	C. Villegas	13.0	M. O'Meara	5.96	J. Kelly	4.98
5.	B. Heintz	12.8	B. Tway	5.83	J.-M. Olazabal	4.95
6.	B. Faxon	12.7	S. Elkington	5.75	T. Byrum	4.73
7.	D. Duval	12.5	T. Petrovic	5.75	S. Leaney	4.67
8.	G. Ogilvy	12.3	C. Parry	5.69	J. Driscoll	4.64
9.	J. Coceres	12.2	P. Sheehan	5.66	G. Kraft	4.35
10.	S. Appleby	12.1	M. Kuchar	5.64	P. Lonard	4.17
	Average	9.0	Average	4.50	Average	2.10

Still, table 7.5 may give you some idea of who is dropping bombs on the greens. The small sample sizes make these rankings suspect. For example, the top three putters in the percentages of putts made from over 50 feet all made exactly 6 putts from long distance. The differences in percentages result from Wes Short attempting 100 putts from over 50 feet compared to Bernhard Langer's 154 attempts. John Senden also made 6 putts of over 50 feet but did not make the top 10 because he attempted 377 at these distances (the most in the data set). Several of the 205 pros in my five-year lists did not make a putt over 50 feet. Geoff Ogilvy (who ranked all the way up at no. 8 from 25–30 feet) went 0 for 191, the most attempts without a make.

From long distance, the goal even for pros is to limit the damage to two putts. The final two putting lists are for percentages of 3-putts (a small percentage is desired) and 1-putts (where a high percentage is good) in 2008. Surprisingly, four golfers made both lists. Apparently, having your last name rhyme with "okay" makes for success on the greens.

Table 7.6 Top ten percentages of 1-putts and 3-putts, 2008

Rank	1-Putts	(%)	3-Putts	(%)
1.	C. Pavin	44.6	L. Donald	1.51
2.	L. Donald	43.8	L. Mattiace	1.64
3.	P. Harrington	43.6	A. Baddeley	1.75
4.	C. Sullivan	43.5	C. Kanada	1.79
5.	N. Green	43.3	D. Chopra	1.81
6.	B. Tway	43.3	B. Tway	1.91
7.	D. Chopra	43.2	B. Gay	2.01
8.	B. Gay	43.1	G. Day	2.01
9.	R. Johnson	42.4	C. Collins	2.10
10.	J. Parnevik	42.3	J. Mallinger	2.16
	Average	38.1	Average	3.43

Daily Grinding

Jack Nicklaus has said that he never missed a putt from inside 5 feet on the last hole of a tournament.[3] He did, of course, miss important putts, but in his mind *he* was never the one who was at fault. It certainly appeared to us fans that Nicklaus never missed a crucial putt. There is evidence that touring professionals as a group have the same mindset.

Figure 7.4 shows percentages of putts made from different distances in 2007. In (a), shots are broken down by par and birdie, in (b), by par and bogey. The actual percentages for distances up to 10 feet are given in table 7.2. The corresponding values for distances up to 15 feet are given in appendix A, table A7.1.

The results are remarkable. For every single distance, the percentage of putts made is higher for par than for birdie. Significantly higher.[4] For every distance up to 9 feet, the percentage of bogey putts made is even higher. Among conceivable explanations, the differences are small enough that they could be due to extra knowledge gained from watching birdie putts slide by the hole. Phil Mickelson is especially blatant about watching the break as the

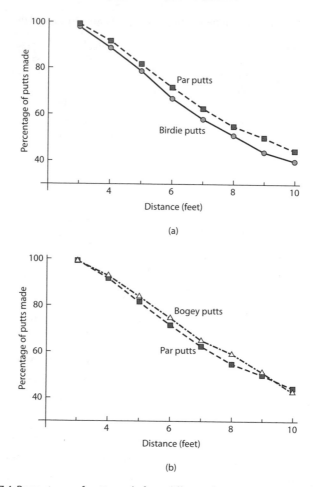

Figure 7.4 Percentages of putts made from different distances when putting for (a) birdie and par and (b) bogey and par

ball goes past. This explanation seems to be borne out in figure 7.5, which shows percentages of putts made from different distances, broken down by whether the putt was a first or second putt. The corresponding values for distances up to 15 feet are given in appendix A, table A7.2.

Again, the comparison is clear. At every distance, more second putts are made than first putts. The differences are larger this time.

This result is at least partly due to small sample sizes. There were only 112 15-foot second putts attempted in 2007, with 52 of them made. (Personally, I feel better about my own putting knowing that there were this many bad first putts.) The improved "read" from watching the first putt likely increases the second-putt percentages.

There is more. The graph in figure 7.6 shows percentages of *first* putts made from different distances, broken down by whether the putt was for birdie or par. The corresponding values for distances up to 15 feet are given in appendix A, table A7.3. The differences persist. Even for the first putts taken by players, the percentage of makes is higher when putting for par. So it is not just personal experience on the green that explains the difference. It could be that watching a chip roll on the green is a benefit. However, the best explanation I can offer is that par putts and second putts are more important, in psychological if not numerical value. Nobody wants to 3-putt or make a bogey or double-bogey. The pros may grind a little harder on bogey putts, par putts, and second putts, with the result that they make slightly higher percentages.

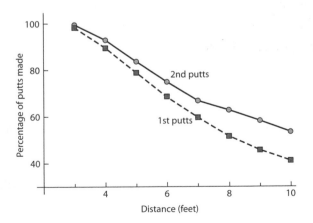

Figure 7.5 Percentages of putts made from different distances, first putt versus second putt

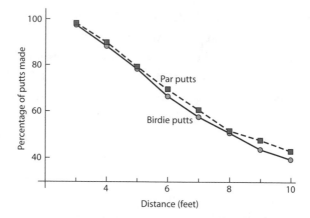

Figure 7.6 Percentages of first putts made from different distances when putting for birdie and par

Chip Sullivan, a top club professional with PGA Tour experience, is not surprised by the difference in percentages. According to Chip, it is less about grinding or focus and more about a total mindset: "It's great to make a birdie putt, but you *absolutely have to* make the par putt."[5] He says that he sometimes tries to convince himself that a putt for birdie is, instead, for par.

Apparently, even for pros, not all shots are created equal. Perhaps what makes them pros, as distinguished from the army of amateur wannabes, is that they convert the extra pressure into an increased percentage of putts made.

The Back Tee: Of Great Significance

From 0–3 feet, the pros made 97.9% of their birdie putts, 99.2% of their par putts, and 99.3% of their bogey putts. Are these differences significant? An increase from 98% to 99% seems dramatic, but is the change from 99.2% to 99.3% worth mentioning? Statistics gives us a way of objectively measuring the likelihood that such differences are purely random.

The conclusions that are drawn from any statistical analysis depend on the assumptions made. Here, I assume that the putts are Bernoulli trials. This means that every par putt from 0–3 feet by every player is an independent event with the same probability p of being made. Every birdie putt from 0–3 feet is also an independent event, with probability b of being made. The question is whether the evidence points to p and b being different.

If you were under the misconception that I am a professional golfer, you could watch me play a few holes and decide otherwise. After each smothered iron and chunked chip, you might think, "It's not likely that a pro would do that." Eventually, enough of these unlikely events would accumulate that you would conclude that this isn't a pro having a bad day; this guy is just not a pro. Similarly, we assume (the null hypothesis) that the probabilities p and b are equal and then compute how likely or unlikely the observed data are. If the actual results are unlikely enough, we conclude that the probabilities are *not* equal. The assumptions made allow us to compute probabilities to precisely quantify "likely" and "unlikely."

To compare par and birdie putts from 0–3 feet, we use the following data: 114,476 out of $n_1 = 115,347$ par putts were made (a success rate of $p_1 = 0.992449$) and 8,962 out of $n_2 = 9,150$ birdie putts were made (a success rate of $p_2 = 0.979454$). The common probability of making a putt is estimated as $\hat{p} = \frac{114,476 + 8,962}{115,347 + 9,150} = 0.991494$, with a standard error of $SE = \sqrt{\hat{p}(1-\hat{p})}\sqrt{1/n_1 + 1/n_2}$. The z-score for the data is $\frac{p_1 - p_2}{SE} \approx 13$.

The z-score is associated with a probability. In this case, the probability of being at least 13 standard deviations above the mean is essentially zero and much less than 0.01. If the probability is less than 0.01, we say that the result is significant at the $\alpha = .01$ level and reject the null hypothesis. In other words, the difference in performance for par versus birdie is too unlikely to be attributable to chance. There is enough evidence to conclude (with almost no doubt) that the pros make significantly more putts from 0–3 feet for par than for birdie.

What about the difference between par and bogey putts from 0–3 feet? The difference between 99.2% and 99.3% does not seem impressive, but is it significant? Repeating the above calculations with $n_2 = 34{,}058$ and $p_2 = 0.993452$ (the pros made 33,835 out of 34,058 bogey putts from 0–3 feet), we get a z-value of about 1.9, with an associated probability of 0.028. The probability is larger than 0.01, so we do not find the result significant at the $\alpha = .01$ level. However, the probability is less than 0.05, so the result is significant at the $\alpha = .05$ level. If we are willing to accept a 5% chance of being wrong, we can reject the null hypothesis and conclude that the pros make significantly more 0- to 3-foot putts for bogey than for par. It turns out that 99.3% can be significantly larger than 99.2%.

Contrast this with the following example that 44% is not significantly higher than 42%. For putts from 9–10 feet, the pros made 1,721 out of 3,869 (44.48%) putts for par and 166 out of 389 (42.67%) putts for bogey. The z-value here is 0.684 with an associated probability of 0.248, well above the usual significance levels of 0.05 or 0.01. The difference between this example and the previous example is the much smaller number of putts involved from 9–10 feet. A handful of lucky putts can make the difference between 42% and 44%. This is one of the few cases in which a higher percentage of par putts than bogey putts is made but the number of putts is not large enough to rule out randomness as the cause.

A nice way of visualizing significance of differences is to use error bars. Here, we compute a standard error that quantifies the accuracy of a measurement. Instead of simply plotting the corresponding data point, we frame the point with error bars that vertically add and subtract the standard error. If error bars from two measurements do not overlap, we have visual evidence that the differences in values are significant and are not likely to be due to random fluctuations.

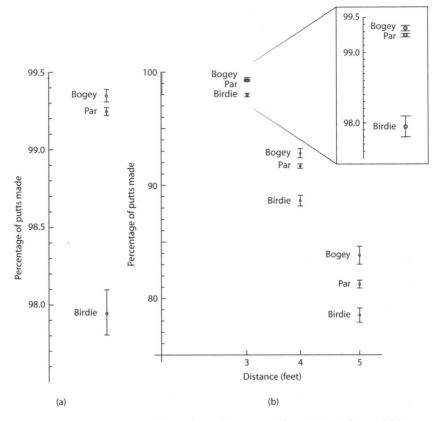

Figure 7.7 Percentages with error bars of putts made from (a) 0 to 3 feet and (b) 3, 4, and 5 feet

Figure 7.7a shows the percentages for made putts from 0–3 feet with error bars. The birdie bar is well below the other two bars, and we saw that there is no question that the birdie percentage is significantly lower than the par percentage. The par and bogey bars are close together but do not overlap. The calculations showed that these percentages are significantly different at the $\alpha = .05$ level but not at the $\alpha = .01$ level. Also, notice that the error bars are much larger for birdie putts (small sample size) than for par putts (large sample size). Thus, error bars contain good information.

In Figure 7.7b, percentages of made putts for 3, 4, and 5 feet are shown on the same graph. At this scale, it is very hard to discern that the bogey and par bars are separate for 0- to 3-foot putts. However, it is apparent that the percentages at distances of 4 and 5 feet are significantly different, with birdie percentages being less than par percentages, which are less than bogey percentages.

For almost all distances less than 10 feet, the differences in made putt percentages are significant when comparing putts for birdie, putts for par, and putts for bogey. In general, the pros do make a higher percentage of their putts for par than for birdie and a higher percentage of their putts for bogey than for par.

EIGHT

Chips and Flops
Short Game Statistics

You create your own luck by the way you play. There is no
such thing as bad luck.

—*Greg Norman*

Greg Norman was in control of the 1986 PGA Championship. He
and Bob Tway were tied for the lead on the 72nd hole, in what
amounted to a sudden-death playoff. Tway had played his second
shot on the par 4, gouging a 9-iron out of heavy rough and short-
siding himself in a bunker. Norman's approach shot spun back
into the fringe 25 feet below the hole, in position for an easy par
that could win the tournament. And then . . . Tway's bunker shot
came out cleanly but appeared to be headed well past the hole
until it rattled the pin and dropped in the hole for a miraculous
birdie that Norman could not match. Unbelievably, this scenario
repeated itself in the very next major, when Larry Mize holed a
140-foot pitch in a playoff with Norman to win the 1987 Masters.[1]
While Norman never did win the Masters, "the Shark" did win the
British Open twice; he also held the 54-hole lead and finished third
at age 53 in the 2008 British Open.

In this chapter, we will look at PGA statistics for various shots
from inside 50 yards. We will see whether, 20 years later, Bob Tway
is unusually skilled from bunkers and Larry Mize is a wizard with
chips.

Fringe Benefits

For official PGA statistics a putt is a stroke taken from somewhere on the green. Often, strokes are taken with a putter from just off the green, but these are not officially classified as putts. Unfortunately, the ShotLink data does not record the club used on a shot, so we do not know whether or not a putter was used from just off the green. Figure 8.1a shows the average number of strokes needed to hole out when starting in the fringe from 12 to 20 feet, along with the curve $y = 0.88 + 0.337 \ln(x - 2)$. Recall that this curve closely matches the average number of putts from different distances. The performance from the fringe is noticeably worse than from the green, where the data closely match the curve.

Figure 8.1b shows the average number of strokes from the fringe from 80 to 100 feet, along with the curve for putting averages. From the longer distances, the fringe performance seems to be very close

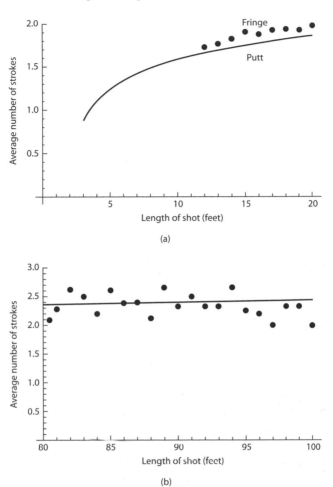

Figure 8.1 Average number of strokes needed when starting in the fringe from distances ranging from (a) 3–20 feet and (b) 80–100 feet. The points represent averages from the fringe, the curve fits averages for putts.

to the averages achieved from the green. A possible conclusion is that hitting from the fringe decreases the pros' chance of holing a short shot but does not strongly affect their ability to lag close to the hole from longer distances.

139

Scrambling on My Mind

The PGA Tour keeps a scrambling statistic to measure players' abilities to save par. The general success rate for scrambling of any sort from 2004 to 2008 was 56%. For example, out of the 120,882 times that a green was missed in 2007, the golfers made par or better 67,560 times, or about 56% of the time. Even the pros sometimes have to get up-and-down (one shot "up" on the green, one shot "down" into the hole) for bogey, so this percentage underestimates the actual success rate for getting up-and-down.

While scrambling rates may be interesting, the focus here is on individual skills. In particular, I want to isolate chipping ability from putting ability. Unfortunately, the data set does not classify which of the many possible types of short shots (chip shot, pitch, flop shot, putt) the player attempted. We can analyze only the result of the shot. For shots taken from off the green, the goal is to get close to the hole; like many painfully obvious statements, this one has some ramifications. In particular, golfers often talk about getting up-and-down. However, an up-and-down consists of two distinct shots, the chip (or pitch or flop) and the putt. The purpose of the chip is to get close, and the purpose of the putt is to go in. So, it makes sense to measure putting by percentage made or average number of putts, while an appropriate measure for chip shots is remaining distance to the hole after the shot.

Getting Close

In 2008, 225 shots were hit from the fairway from distances between 49 and 50 yards from the hole. These shots stopped a total distance of 40,134 inches from the hole. This means that, from 50 yards out, the pros hit the ball an average of $\frac{40,134}{225} = 178.4$ inches, or about 14.9 feet, from the hole. Figure 8.2 shows the results of analogous calculations for distances ranging from 4 yards to 50 yards.

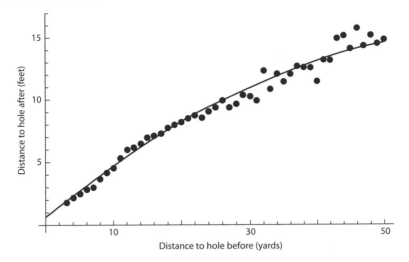

Figure 8.2 Average distance to the hole after shots from the fairway, 4–50 yards

The points form a fairly smooth curve, looking like part of a downward parabola. The curve $y = -0.0033(x - 3)^2 + .43(x - 3) + 1.57$ is superimposed on the data points.

The curve illustrates the obvious rule that the closer you are to the hole, the closer you tend to hit your approach. The downward trend indicates that the decrease in accuracy lessens as you approach 50 yards. For example, the loss in accuracy changing from 10 to 20 yards out is 3.79 feet, which is almost twice the loss in accuracy backing up from 40 to 50 yards (1.98 feet). (See table 8.1.) A likely explanation is that the ability to spin the ball better from the longer distances partially compensates for the increased distance of the shot.

Lies and Damn Lies

Figure 8.2 summarizes all shots from the given distances from the fairway in 2008. Obviously, the approaches will not be as accurate from other lies, such as the rough. To show how much missing the

Table 8.1 Differences in accuracy of approach shots from various distances and lies

	Hitting from				
Lie	10 yds out (ft)	20 yds out (ft)	30 yds out (ft)	40 yds out (ft)	50 yds out (ft)
Fairway	4.2	8.0	10.4	12.6	14.6
Intermediate rough	4.8	8.4	9.4	16.1	17.9
Primary rough	6.9	11.7	15.8	20.8	22.5

fairway costs pro golfers, table 8.1 lists averages from several distances for shots from the fairway, intermediate rough, and primary rough. The intermediate rough values tend to be much closer to the fairway values than to the primary rough values. Far more balls in the data set are marked as being in the primary rough than in the intermediate rough, so the intermediate rough data are spotty and, for this reason, mostly ignored in what follows.

The data do not account for trees at all. The very large values from the primary rough could be due partly to having to chip sideways to avoid trees. The evidence seems to show that being in the intermediate rough does not cost the pros much but that being in the primary rough is indeed a disadvantage. Figure 8.3 shows the averages from the rough for distances ranging from 4 yards to 50 yards, with the curve $y = -0.0037(x-3)^2 + .59(x-3) + 2.84$ superimposed.

At first glance, figures 8.3 and 8.2 look very similar. They show the same characteristics of rising from left to right but gently curving along a downward parabola. The most important difference in the figures is the vertical scale. Shots from the rough in figure 8.3 end up farther from the hole than the equivalent shots from the fairway in figure 8.2. To make this comparison more obvious, the two sets of data are combined in figure 8.4.

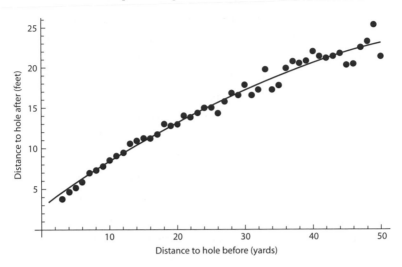

Figure 8.3 Average distance to the hole after shots from the rough, 4–50 yards

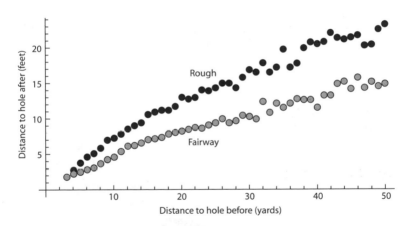

Figure 8.4 Average distance to the hole after shots from the fairway and the rough, 4–50 yards

Playing in the Sand

For casual golfers, sand bunkers are terrifying hazards to be avoided at all costs. The pros are so good from the sand that they often think of bunkers as safe havens from gnarly rough. How good are they?

Out of 30,288 trips to greenside bunkers in 2007, the pros recorded 14,568 saves, a 48% recovery rate. Of course, a sand save can be as much a function of great putting as great bunker work. To isolate ability to hit out of the sand, we look at how close to the hole pros are getting their sand shots. For bunker shots that start between 20 feet and 150 feet from the pin, the average shot finished 12.35 feet from the hole. Figure 8.5 shows average distances from 20 to 150 feet, with the curve $y = 0.0008(x - 20)^2 + 0.035(x - 20) + 9.13$ superimposed. Broken down by 30-foot distance ranges, table 8.2 shows the average distances to the hole after bunker shots of different lengths over the five-year period from 2004 to 2008.

Even from 100 feet, the pros are giving themselves a decent putt at par.

Do the pros do better out of the sand than out of the rough? The averages computed here cannot completely answer the question, but they do provide a starting point for discussion. For shots from both the rough and sand, smooth curves can be fit to the data. The two curves, shown in figure 8.6, demonstrate different characteris-

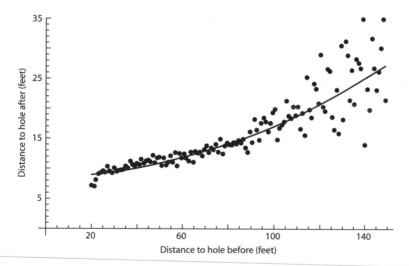

Figure 8.5 Average distance to the hole in feet after shots from the bunker, 20–150 feet

Table 8.2 Average distances to the hole after
bunker shots of various lengths, 2004–2008

Bunker shots	Average distance to hole (ft)
From 0–30 ft	9.52
From 30–60 ft	11.06
From 60–90 ft	13.10
From 90–120 ft	17.44
From 120–150 ft	23.49

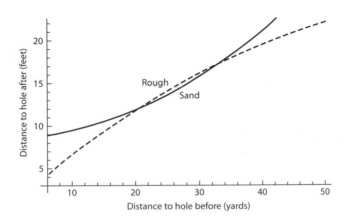

Figure 8.6 Fitted curves for average distance to the hole from the rough and sand,
7–50 yards

tics. For sand shots, the parabola curves up, indicating that longer
bunker shots are much harder than shorter shots. By contrast, the
curve for shots from the rough is a downward parabola, indicating
that longer shots from the rough are not that much harder. This
should match your experience on the course. From 20 yards to
35 yards, the average distances are about the same, with the pros
doing slightly better from the sand than from the rough.

Individual Stats

Having seen Tour averages for various short-game statistics, we
now want to see which individuals perform the best. Imagine a

pro golfer needing to get up-and-down at a crucial moment in a tournament. Who would be most likely to get the ball close? See if your choice shows up in the tables that follow.

We start with the 2009 top tens for average approach distance for shots made from 4–50 yards, from both the fairway and rough. You may notice that nobody made both top tens. There is a problem with the statistic shown in table 8.3: averaging all of the shots from 4 to 50 yards could be unfair. If one player was always less than 10 yards away and another player was always more than 40 yards away, a comparison of the averages is misleading. Because the distance range of 4–50 yards is broad, we need a way to adjust for the actual distances faced by each player. The solution depends on a technique that I use repeatedly when determining player ratings. We have average approach distances for 4 yards, 5 yards, and so on. (Figure 8.2 shows these averages.) For each shot from the range 4–50 yards, compare the distance of the player's approach to the Tour

Table 8.3 Top ten average approach distances for shots made from 4–50 yards, 2009

	Fairway		Rough	
Rank	Player	Distance (ft)	Player	Distance (ft)
1.	O. Uresti	5.25	F. Couples	7.90
2.	B. Estes	5.27	T. Immelman	8.21
3.	E. Els	5.47	C. Pavin	8.29
4.	A. Kim	5.60	G. Day	8.33
5.	T. Woods	5.65	S. Sterling	8.43
6.	J. Furyk	5.74	M. Brooks	8.44
7.	B. Gay	5.75	F. Lickliter	8.45
8.	S. Garcia	5.83	F. Jacobson	8.51
9.	S. Stricker	5.86	T. Clark	8.61
10.	J. Kaye	5.92	D. Stiles	8.66
	Average	7.72	Average	11.41

average from that distance. For example, an approach played to 6 feet from 10 yards away is 1.79 feet worse than the Tour average of 4.21. However, an approach played to 7 feet from 50 yards away is 7.55 feet better than the Tour average of 14.55. The second approach is farther from the hole but is a more impressive shot compared to the Tour average. For the two shots, the player would be $-1.79 + 7.55 = 5.76$ feet better than average. On average, the player is $\frac{5.76}{2} = 2.88$ feet per shot better than average.

In table 8.4, we see the top tens from the fairway and rough for 2009, compared to tour averages. Most of the names in the top tens are the same as in table 8.3, but considerable shuffling has occurred. Tiger Woods now heads the list from the fairway; the rating of 2.47 means that Tiger's average shot from the fairway (4–50 yards away) finished 2.47 feet closer to the hole than the average Tour shot from the same distance. Because Woods moved up the list and Omar Uresti dropped down three slots, we can infer that Tiger played from longer distances throughout the year than did

Table 8.4 Top ten average approach distances for shots from 4–50 yards, compared to Tour averages from the same distances, in 2009

	Fairway		Rough	
Rank	Player	Distance (ft)	Player	Distance (ft)
1.	T. Woods	2.47	C. Pavin	2.66
2.	B. Estes	2.31	D. Hart	2.65
3.	E. Els	2.23	G. Day	2.56
4.	O. Uresti	2.17	F. Jacobson	2.34
5.	B. Gay	1.98	S. Sterling	2.26
6.	T. Lehman	1.96	C. Wittenberg	2.19
7.	L. Donald	1.89	D. Stiles	2.07
8.	J. Leonard	1.88	M. Brooks	2.07
9.	J. Furyk	1.82	F. Couples	1.99
10.	A. Kim	1.81	T. Wilkinson	1.97

Omar. The top tens for 2006–2008 are given in Appendix A, tables A8.1a–A8.1c.

One conclusion that can be drawn from the data in these tables is that there is little consistency from one year to the next. It is also interesting that the players who are strong from the rough are not necessarily the ones who make the top ten from the fairway. The need to accurately judge the effect of the lie in the rough makes it a different type of shot.

The same analysis can be applied to sand shots. Table 8.5 shows the top tens in bunker play for 2009, both in terms of raw distance and distance better than Tour average. There is not a large difference in the two rankings. Table 8.6 shows five-year totals, broken down by distance. About 55% of the bunker shots in 2004–2008 were taken from the 30–60 foot range, and 27% were taken from the 60–90 foot range. More lists are given in Appendix A, table A8.2. Based on the data in these tables, we conclude that Mark Wilson, Mike Donald, Mike Weir, Nick Price, Jeff Sluman, Omar Uresti, and Phil Mickelson were among the top bunker players in the 2000s.

Table 8.5 Top ten average distances from bunkers, raw and compared to Tour averages, 2009

Rank	Raw data	(ft)	Compared to Tour average	(ft)
1.	R. Allenby	7.55	R. Allenby	4.31
2.	A. Scott	8.76	J. Oh	2.93
3.	K. Na	8.92	M. Weir	2.88
4.	M. Weir	9.00	A. Scott	2.84
5.	S. Micheel	9.02	N. Green	2.83
6.	N. Green	9.06	N. O'Hern	2.69
7.	K. Stanley	9.08	C. Riley	2.66
8.	O. Uresti	9.11	G. Day	2.60
9.	N. O'Hern	9.24	B. Molder	2.47
10.	G. Day	9.28	K. Na	2.40

Table 8.6 Top ten average approach distances for shots from bunkers, 2004–2008

Rank	All distances	(ft)	30–60 feet	(ft)	60–90 feet	(ft)
1.	M. Wilson	9.14	M. Wilson	7.89	N. Price	8.80
2.	L. Donald	9.33	M. Weir	8.48	D. Toms	8.83
3.	M. Weir	9.62	L. Donald	8.55	J. Sluman	9.29
4.	C. Pavin	9.63	C. Wi	8.62	T. Clark	9.58
5.	N. Price	9.69	N. Price	8.76	S. Appleby	9.67
6.	O. Uresti	9.71	O. Uresti	8.79	J. Coceres	9.74
7.	J. Sluman	9.89	C. Riley	8.87	V. Singh	9.84
8.	P. Mickelson	10.07	C. Pettersson	8.93	P. Mickelson	10.04
9.	G. Ogilvy	10.11	K. Sutherland	8.97	M. Kuchar	10.21
10.	R. Pampling	10.17	R. Sabbatini	8.98	L. Donald	10.22
	Average	12.25	Average	11.06	Average	13.10

In the introduction to the chapter, I referred to Greg Norman's difficulties in the majors in 1986 and 1987. ShotLink data are not available for those years, so there are not valid conclusions to be drawn, but I thought it would be interesting to see where Larry Mize and Bob Tway ranked in the short-game stats. In 2006, from 4–50 yards in the fairway, Larry Mize ranked 68th out of 230 golfers, well above average. In 2007, Bob Tway ranked 190th out of 230 golfers in bunker play, averaging nearly 3 feet per shot farther from the hole than the average PGA golfer. Nevertheless, in 1986 he made the one that mattered the most.

The Back Tee: Round Off the Usual Suspect

The ShotLink data sets are stunningly detailed but not complete. For example, in the first round of the 2008 Barclays tournament, on the 5th hole, Steve Stricker faced a second shot of 2,312 inches (about 64 yards). The ball traveled a distance of 2,293 inches, coming to rest 59 inches (about 5 feet) from the hole. (In case you're wondering, Stricker made the putt for birdie. The time of day was 4:56 p.m.)

Stricker's shot finished 5 feet away, but was it long or short? Left or right? The data set does not record the answers, but the first question is easy enough to resolve. The ball started 2,312 inches from the hole and flew only 2,293 inches, so the fact that $2,293 < 2,312$ tells us that the shot came up short. We can also determine how far off-line the shot was, but whether the shot missed left or right will remain unknown. Start with the fact that the ball stopped 59 inches from the hole. The ball must be on a circle of radius 59 inches centered at the hole. Similarly, the ball traveled 2,293 inches from its starting point, so the ball also lies on a circle of radius 2,293 inches centered at Steve Stricker. As seen in figure 8.7 (representing an overhead view of the hole with Stricker standing somewhere below the bottom of the page), the two circles intersect in two locations.

The ball must be located at one of the two intersection points in figure 8.7. The possible locations of the ball can be found using some algebra. Locate the center of the hole at $(0, 0)$, and let the distance to the hole after the shot be H. An equation of the circle of radius H centered at the origin is $x^2 + y^2 = H^2$. Place the golfer d inches away at $(0, -d)$, and let B be the distance the ball travels in

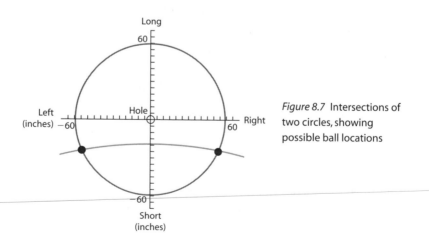

Figure 8.7 Intersections of two circles, showing possible ball locations

inches. An equation of the circle of radius B centered at $(0, -d)$ is $x^2 + (y + d)^2 = B^2$. To find the intersections of the circles, expand $(y + d)^2$ and subtract one equation from the other. You should find that

$$y = \frac{B^2 - H^2 - d^2}{2d},$$

and then $x = \pm\sqrt{H^2 - y^2}$. The y-value tells whether the shot is long ($y > 0$) or short ($y < 0$). We can't distinguish whether x is positive or negative, so we can't determine whether the ball is to the left or right. In what follows, we use both possibilities.

Figure 8.8 shows the computed stopping points for approximately 2,000 shots from the fairway from 10–20 yards out. The stopping points range from about 20 feet short to 20 feet long, with a fairly symmetric distribution. (The left-right symmetry is artificial, as I have plotted each point twice—once to the right and once to the left.) The shots were up to 8 feet off-line. Two features stand out to me in figure 8.8. First, many of the shots seem to have been

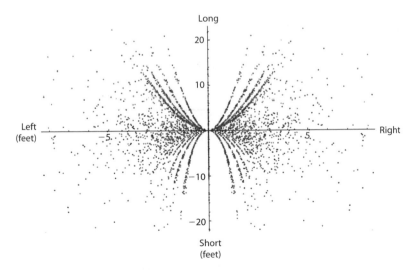

Figure 8.8 Computed locations of shots from the fairway, 10–20 yards away

exactly on-line. Far more worrisome are the bands of points branching out like parabolas. If you've read too many Knights Templar conspiracy novels lately, you might be trying to decipher the secret message embedded in the PGA Tour data sets. As we will see, the real message is to beware of data that are rounded off.

In our example, Steve Stricker's approach finished 59 inches from the hole. It is unlikely that the ball was exactly 59 inches away, but rounding off 58.7 inches to 59 inches seems harmless enough. In particular, rounding off to the inch couldn't produce the gaps in figure 8.8, could it? After all, at $y = 10$, there are no points plotted for x-values between 0 inches and nearly 1 foot.

Surprisingly, the gaps in figure 8.8 are caused by rounding the data to the inch.[2] To illustrate this, I took the data used to produce figure 8.8 and "unrounded" it by adding random decimals. The result in figure 8.9 shows that the gaps have been mostly filled in. The stopping points for shots from the fairway from 10–20 yards out should look like those in figure 8.9. Rounding off the location

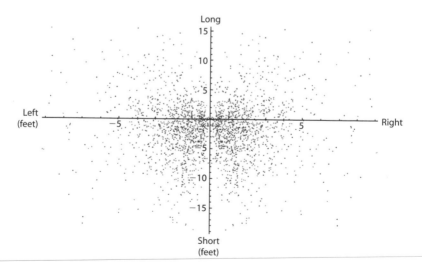

Figure 8.9 Locations of shots from the fairway, 10–20 yards away, with unrounded data

of each shot to the inch has the surprising effect of producing the bands seen in figure 8.8. Simple mathematical processes like rounding can have unexpected consequences.[3]

This technique yields some interesting conclusions, which are presented in the next chapter, where we look at longer approach shots.

NINE

Iron Byron
Approach Shot Statistics

I play golf with friends sometimes, but there are never friendly games.

—*Ben Hogan*

The true story told in Mark Frost's *The Match* starts with wealthy businessman Eddie Lowery bragging at a party on the eve of the 1956 Bing Crosby Clambake. Lowery sponsored two amateur golfers (Ken Venturi and Harvie Ward) who, he claimed, could beat any twosome in a best-ball match. Another businessman, George Coleman, agreed to a big-money bet pitting Lowery's amateurs against two of Coleman's friends. The friends turned out to be Ben Hogan and Byron Nelson. Frost weaves details of the terrific match with the dramatic life stories of the four golfers involved.

The match was played at the spectacular and difficult Cypress Point on the Monterey Peninsula in California. Hogan holed out an 85-yard wedge for eagle on the 10th on his way to a course record of 63. Venturi shot 65, while Nelson and Ward were "only" 5 under par with 67s. Hogan and Nelson had a best-ball score of 57 to win by 1.

One of the most dramatic moments of the match came on the par-3 16th hole, one of the most famous holes in golf featuring a 220-yard carry over water to a green perched on a small peninsula reaching into the Pacific Ocean. Hogan and other top golfers often chose to play their tee shots safely to fairway, 40–60 yards short and

left of the green. On this day, all four golfers hit drivers. Hogan's drive found the green; Nelson hit an incredible drive that landed 6 feet from the hole; Venturi hit the green; and, hitting last under all the pressure, Ward launched a frozen rope that landed 5 feet from the hole. Nelson and Ward both made their birdies, and the match blazed on.

An interesting historical footnote to this match is the important role that caddying played. Eddie Lowery, the instigator of the match, had achieved fame as a 10-year-old boy: he was Francis Ouimet's feisty caddie in the 1913 U.S. Open, which was dramatized in the movie *The Greatest Game Ever Played*.[1] Hogan and Nelson both learned the game as caddies. Amazingly, Nelson and Hogan were caddies at the same country club, Glen Garden Country Club in Fort Worth, Texas. In fact, in 1927 Nelson won the caddie championship at Glen Garden by beating Hogan in a playoff.[2]

Their careers remained intertwined until Nelson's early retirement in 1946, only one year after his remarkable eleven straight victories.

In their prime, Hogan and Nelson were considered the best iron players in golf. Almost all golf pro shops display the famous picture of Hogan hitting a 1-iron to the 18th green to win the 1950 U.S. Open. Nelson was immortalized by the USGA when it nicknamed its mechanical swing machine "Iron Byron." In this chapter, we will see which modern players are especially accurate with irons as we examine approach shot statistics for distances ranging from 50 to 250 yards.

If 6 Was 9

The above story about 220-yard drives may seem quaint. Modern technology has improved distance and accuracy with drivers and other clubs; however, unlike Hogan, nobody hits a 1-iron anymore. The main reason for the loss of this ability is that the defining properties of irons have changed. If you hit your 6-iron about 10 yards farther now than you did 20 years ago, it may be due to better swing mechanics but is probably because your new 6-iron is the same as your old 5-iron. Primarily for marketing reasons ("Gain 10 yards with our revolutionary new club design!"), the average loft of a 6-iron decreased from 36° in the 1960s to 34° in the 1980s to 30° in 2000.[3] The 30° 6-iron of 2000 is almost identical to the 30° 5-iron of the 1980s.

Going back a little further in time, we could speculate that Ben Hogan's 30° club in 1950 may have had a "4" stamped on it. Thus, Hogan's famous 1-iron may have played like a modern 3-iron. By analogy, a modern 1-iron would be equivalent to an old "−1"-iron. Even Hogan would have struggled with such a thing.

Before looking at the approach shot statistics, let's pause to explore the difference that the club loft makes.

The Compleat Angler

How many different swings do you have? Do you have a different swing for each iron? Although some adjustments in swing plane and length of backswing are common, the design of golf clubs allows you to keep essentially the same swing for each iron. As you move from 5-iron to 9-iron, the clubs get progressively shorter and more lofted (i.e., the clubhead points higher above the ground). Both of these changes affect the height and distance of a solid shot, giving you the desired 10-yard gap between irons with little adjustment to your swing.

In table 9.1, a ball is given a fixed launch speed of 120 mph and backspin of 4,200 rpm, and the carry distance is computed. The angles shown correspond to common lofts of the 5-iron through the 9-iron. Notice that the yardage differences from club to club become increasingly larger, with the change in angle from 28° to 31° taking off only about 5 yards, while the change from 37° to 40° takes off over 8 yards. How do you get to a constant 10-yard gap between clubs?

The assumption of equal spin rate and launch speed for each iron, made for the data in table 9.1, is not at all realistic. Spin is a product of the contact between the club and ball. As the club slides under the ball and digs into the ground, the ball rolls up the

Table 9.1 Carry distances for a launch speed of 120 mph with 4,200-rpm backspin, various angles

Launch angle (deg)	Distance (yd)
28	193.3
31	188.6
34	182.6
37	175.4
40	166.9

clubface. The more rolling there is, the more spin the ball has. This is why it is sometimes difficult to get spin on the ball out of the rough: the grass can get in the way and prevent the ball from staying in contact with the club long enough to generate maximum spin. This is also why using "the grease" (e.g., rubbing WD-40 on the clubface) on a driver can reduce a hook or slice: a slicker clubface imparts less spin. Less sidespin, in turn, produces less sideways movement. (However, less backspin also produces less height and possibly less distance.)

Think of hitting a ping pong ball with a paddle. If you hit it "square," or perpendicular, as in figure 9.1a, you will get maximum ball speed and little or no spin. If you chop down with an angled paddle, as in figure 9.1b, you will gain spin but lose ball speed. A similar effect occurs on a golf shot. For a 3-iron, the ball flattens against the clubface at impact and then rebounds off the club at high speed, but the lack of loft allows for little rolling and produces a low spin rate. For a 9-iron, the extra loft produces more of the up the clubface chop in figure 9.1b, and the ball leaves with more spin and less speed. Changes in club loft are automatically accompanied by changes in spin rate and ball speed.

There is another large error in the assumptions behind the table 9.1 data. We assumed that the launch angle of the ball is the same as the loft of the club. The interaction between ball and club also affects the launch angle. There are two sources of energy loss as the ball moves up the clubface. Friction between club and ball slows the ball's climb up the clubface, and energy is lost as the ball

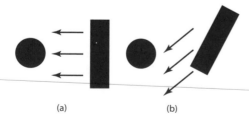

Figure 9.1 Square hit versus angled hit: (a) more speed, less spin; (b) less speed, more spin

(a)　　　　　(b)

Table 9.2 Launch angle, ball speed, spin rate, and carry distance for clubs of different loft, adjusted for different club lengths

Club loft (deg)	Launch angle (deg)	Ball speed (mph)	Spin rate (rpm)	Carry distance (yd)	Adjusted (yd)
28	24.2	120	4230	197.4	197.4
31	26.5	118	4637	189.6	187.2
34	28.7	115	5034	180.7	176.1
37	30.8	112	5419	171.1	164.6
40	32.7	109	5788	161.0	153.3

compresses against the club. The net result is that the ball leaves the club at a smaller angle than the loft of the club. Equations for the actual angles are adapted from Jorgensen's *The Physics of Golf*.[4]

The more realistic comparison in table 9.2 adjusts the spin rate, launch speed, and launch angle for each club. There is now about a 9-yard difference in distance between irons. In this table, ball speed is measured in mph, spin rate in rpm, and carry distance in yards. A final adjustment shown in the last column is to modify the ball speed to take into account the different lengths of the clubs. As you can see, this has a significant effect on distance and produces a 10- to 11-yard gap in carry distance between clubs.

If you do not have a 10-yard gap between clubs, the problem may be with the clubs. If the manufacturing process was imperfect or the club has been slightly bent, a club may not have the loft that it is supposed to. An audit and adjustment to get your irons properly lofted might improve your game substantially.

Tour Statistics

The pros, presumably, have their clubs properly fitted and adjusted. How much control do they have over their irons? We first look at tour averages using the ShotLink data from 2007. Other years show similar averages.

In 2007, 840 shots were hit from the fairway at distances between 99 and 100 yards from the hole. These shots stopped a total distance of 190,179 inches from the hole. This means that, from 100 yards out, the pros hit the ball an average of $\frac{190,179}{840} = 226.4$ inches, or about 18.9 feet, from the hole. Figure 9.2 shows the results of analogous calculations for distances ranging from 4 yards to 300 yards for shots from the fairway.

The points form a fairly smooth curve, but the characteristics of the curve change at about 50 yards. The distance range of 4–50 yards is analyzed in chapter 8. The data points from 50–300 yards seem to curve as part of an upward parabola. This is illustrated in figure 9.3, showing the data from 100–200 yards, with the curve $y = 0.0018(x - 100)^2 + 0.033(x - 100) + 19.9$ superimposed.

This curve illustrates the obvious rule that the closer you are to the hole, the closer you can hit your approach shot. The question is what happens if the distance of the shot doubles. Going from 25 to 50 yards, the pros do not lose that much accuracy. The effects of the increased distance are partially offset by the ability to spin the ball better at 50 yards than at 25 yards. Going from 100 to 200 yards,

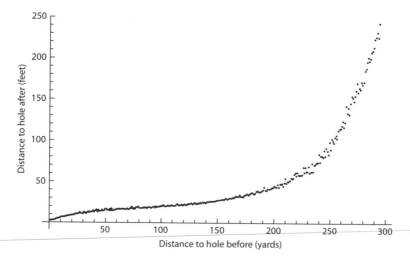

Figure 9.2 Average distance to the hole from the fairway, 4–300 yards

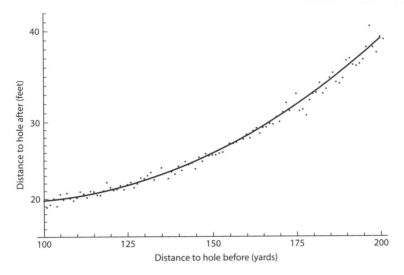

Figure 9.3 Average distance to the hole from the fairway, 100–200 yards

much more accuracy is lost because of the reduced ability to spin the ball well and the increased effect of wind, awkward lies, and so on.

Roughing It

We saw in chapter 8 that being in the rough strongly affects the pros' accuracy from short distances. The next set of figures shows the effect of missing the fairway for longer distances. Table 9.3 lists averages from several distances for shots from the fairway, intermediate rough, and primary rough.

The very large values from the primary rough could be partly due to factors not apparent in the data sets. For example, a ball hit into the rough might also be behind a tree, forcing the player to chip sideways instead of going for the green. There are also situations where the golfer lays up to a favorite distance rather than trying to hit the green. Nevertheless, the evidence is that being in the intermediate rough does not cost the pros much. By contrast, the primary rough seems to be a killer. Figure 9.4 shows the data from all distances ranging from 4 to 300 yards.

Table 9.3 Average distance to the hole from different lies

Lie (yd)	Fairway (ft)	Intermediate rough (ft)	Primary rough (ft)
50	15.3	22.7	24.1
75	16.4	20.1	29.9
100	18.9	20.5	37.2
125	22.2	24.7	49.4
150	25.7	29.9	60.9
175	31.5	39.2	82.4
200	41.0	47.4	135.6
225	57.2	66.7	165.3

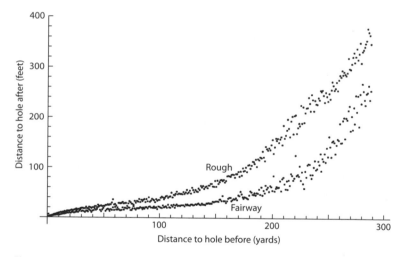

Figure 9.4 Average distance to the hole from different lies, 4–300 yards

Player Statistics

The challenge here is to be informative without being encyclopedic. We can rank players on approach shots from any distance range, from both the fairway and the rough. We can rank players by raw distances or by distances compared to Tour averages. With so many choices, it is difficult to know where to start and when to stop.

My choices are fairly Tiger-centric. Table 9.4 shows approach statistics (from ShotLink data) from the fairway in 2009. The small 25-yard distance intervals eliminate the need to compare to Tour averages. You may wonder why I called these tables "Tiger-centric." Woods made two top tens, but so did Jason Bohn, Brandt Jobe, and Chad Campbell. Rickie Fowler and Tim Clark made three top tens. Tiger has high ranks in all categories (5 out of 6 in the top fifty and all at 71st or better), but there is nothing about the tables that screams "Tiger!" As we will see, in some ways Tiger was by far the best iron player on tour in 2009, but in various categories other players had an extremely good year. It makes sense that no one player could make many of these top ten lists in a given year.

This logic points out the extraordinary nature of Tiger's 2004 to 2008 seasons. In 2008, Tiger was first or second at every distance from 100 yards to 225 yards except 125–150 yards. This is remarkable consistency and dominance. It is true that 2008 was an injury-shortened season for Tiger, so it is wise to wonder whether those rankings are a one-year fluke. Table 9.5 answers that question decisively.

The context for the rankings is that the statistics are for the 230 golfers who played in the most ShotLink events each year. While a rank of 100 may seem unimpressive, it actually represents an above-average performance. Knowing this, we would find a consistent first-place ranking to be outrageous. With the best 230 golfers going at it, we might expect that five or six of them would have a great year from 100–125 yards, a different five or six would have a great year from 125–150 yards, and so on. For one golfer to dominate multiple distances, as Tiger did, for multiple years is stunning. Notice, however, that Tiger's ranking are significantly worse from 125–150 yards than from other distances. Perhaps there is a club that gave Tiger trouble.

Before we look at five-year averages from the different distances, here is one more table from the 2009 season. Given the many

Table 9.4 Top ten average approach distances from fairway, 2009

	Shot from 75–100 yards		Shot from 100–125 yards		Shot from 125–150 yards	
Rank	Player	Distance (ft)	Player	Distance (ft)	Player	Distance (ft)
1.	S. Stricker	13.0	T. Clark	14.7	G. McDowell	18.7
2.	S. Garcia	13.3	C. Wi	16.1	B. Heintz	18.7
3.	R. Fowler	13.4	R. McIlroy	16.3	J. Leonard	18.8
4.	J. Leonard	13.6	R. Fowler	16.5	C. Riley	19.1
5.	N. O'Hern	13.7	L. Donald	16.6	D. Toms	19.2
6.	B. Jobe	14.0	B. Molder	16.7	T. Clark	19.3
7.	C. Villegas	14.1	S. Appleby	16.7	P. Harrington	19.6
8.	B. Van Pelt	14.1	J. Bohn	16.8	B. DeJonge	19.8
9.	D. Toms	14.4	V. Singh	16.9	S. Ames	19.8
10.	G. Chalmers	14.5	K. Stadler	17.0	O. Uresti	19.8
	Average	18.1	Average	20.1	Average	22.9

different players who showed up in table 9.4, is there any way to summarize the statistics from 100–250 yards? Average distance is not reasonable, because the distance gap is too large; comparing a shot from 100 yards to one from 250 yards makes little sense. Instead, let's go back to a previous method of measuring each shot against the Tour average from that distance. In the 2009 season, the average Tiger Woods shot from the fairway finished 6.23 feet closer to the hole than the average Tour shot from that distance. The top ten ranking against the Tour average in 2009 from the fairway at distances of 100–250 yards is shown in table 9.6. Although the top ten lists in table 9.4 don't show it, Tiger Woods clearly had the best year hitting close from the fairway in 2009.

Table 9.7 shows average approach-shot distances for different-length shots for the 2004 through 2008 seasons. The leading averages over the five-year span are significantly higher than the averages for just 2009 (shown in table 9.4). This emphasizes the remarkable nature of all of Tiger's number 1 yearly rankings. Another way

Shot from 150–175 yards		Shot from 175–200 yards		Shot from 200–225 yards	
Player	Distance (ft)	Player	Distance (ft)	Player	Distance (ft)
D. Duval	23.8	E. Romero	26.7	A. Atwal	34.6
T. Woods	23.9	K. Stanley	28.0	B. Jobe	35.3
D. Lee	24.1	F. Couples	28.2	C. Franco	36.1
R. Sabbatini	24.1	C. Campbell	28.6	R. Fowler	36.2
P. Casey	24.2	D. Pride	28.7	S. Ames	36.3
E. Els	24.3	T. Woods	28.8	P. Perez	36.9
J. Bohn	24.4	D. J. Trahan	29.0	R. Gamez	37.1
T. Lehman	24.5	R. Barnes	29.4	T. Ridings	37.2
T. Clark	24.5	R. Beem	29.6	J. Williamson	37.6
C. Campbell	24.6	M. Weir	30.0	J. Holmes	37.9
Average	27.7	Average	36.0	Average	46.7

Table 9.5 Tiger Woods's rankings, by approach distances from fairway, 2004–2008

	Ranking				
Distance (yd)	2004	2005	2006	2007	2008
100–125	1	1	3	3	1
125–150	52	1	44	7	100
150–175	11	51	1	1	1
175–200	6	3	1	1	1
200–225	2	29	1	4	2

of noting Tiger's dominance is to see how closely clustered most of the top players' averages are and then to note the gap from Tiger to number 2.

Second Cut Is the Deepest

We next examine play from the rough for various distance ranges. I have chosen to cut off the lists at 150 yards and to show only the five-year averages.[5] At some point, smart players will start laying up to their favorite distance instead of going for the green. Further, for reasons detailed below, a large sample size is more likely to give meaningful information.

Table 9.6 Top ten average approach distances from fairway, 100–250 yards, compared to Tour average, 2009

Rank	Player	Distance (ft)
1.	Tiger Woods	6.23
2.	Ernie Els	4.28
3.	Tim Clark	3.91
4.	Jason Bohn	3.89
5.	Fred Couples	3.72
6.	Dicky Pride	3.49
7.	Scott Verplank	3.38
8.	Chad Campbell	3.38
9.	Padraig Harrington	3.24
10.	Cameron Beckman	3.16

Table 9.7 Top ten average approach-shot distances from fairway, 2004–2008

Rank	Shot from 75–100 yards		Shot from 100–125 yards		Shot from 125–150 yards	
	Player	Distance (ft)	Player	Distance (ft)	Player	Distance (ft)
1.	S. Stricker	15.1	T. Woods	16.1	S. Verplank	20.8
2.	S. Cink	15.4	S. Verplank	17.6	G. Willis	21.1
3.	S. Verplank	15.4	T. Immelman	18.3	D. Toms	21.1
4.	N. O'Hern	15.5	P. Mickelson	18.3	D. Stiles	21.2
5.	T. Clark	15.6	T. Clark	18.4	B. Gay	21.3
6.	L. Janzen	15.7	J. Leonard	18.5	G. Coles	21.4
7.	J. Leonard	15.7	C. Parry	18.5	P. Mickelson	21.4
8.	M. Gogel	15.7	J. Huston	18.5	T. Woods	21.5
9.	P. Goydos	15.7	S. Stricker	18.6	V. Taylor	21.5
10.	R. Imada	15.8	D. Toms	18.6	S. Cink	21.6
	Average	18.2	Average	20.7	Average	23.8

Interpretations of the rough statistics are, actually, rough. In table 9.8, no player made the top ten from more than one distance. There are numerous obstacles in the rough which make consistent excellence difficult. A horrendous lie or two can significantly raise your average with only a small number of shots. Also, due to trees, water, or other hazards, there may be a shot or two for which the prudent play is a chip out into the fairway. Such a shot may be a smart play, yes, but your average distance to the hole just took a big hit. For these reasons and others, statistics from the rough reflect more than just the ability to get an iron cleanly on the ball.

The Back Tee: Coming Up Short

In chapter 8, we examined a calculation to determine the location of the ball from the golfer's perspective. To be precise, we can determine how far long or short the shot is, and we can determine how far off-line the shot is, but we can't distinguish whether it missed right or left. Here, we use that calculation to learn an interesting fact about where the pros miss.

Shot from 150–175 yards		Shot from 175–200 yards		Shot from 200–225 yards	
Player	Distance (ft)	Player	Distance (ft)	Player	Distance (ft)
T. Woods	24.5	T. Woods	27.2	T. Woods	36.9
A. Cejka	25.9	E. Els	31.1	E. Els	37.5
J. Furyk	25.9	A. Cejka	31.3	C. Kresge	40.7
K. Perry	26.2	S. Garcia	31.8	W. Short	42.1
S. Garcia	26.2	T. Immelman	32.1	H. Slocum	42.2
S. Cink	26.2	H. Slocum	32.2	T. Clark	42.2
S. Ames	26.4	R. Garrigus	32.3	T. Immelman	42.4
R. Goosen	26.4	L. Glover	32.3	B. Jobe	42.5
T. Lehman	26.5	S. Ames	32.3	B. Watson	42.8
S. Verplank	26.5	C. Pettersson	32.4	J. Maggert	42.8
Average	29.0	Average	36.1	Average	49.1

Table 9.8 Top ten average approach distances from rough, 2004–2008

	Approach shot made from 75–100 yards		Approach shot made from 100–125 yards		Approach shot made from 125–150 yards	
Rank	Player	Distance (ft)	Player	Distance (ft)	Player	Distance (ft)
1.	P. Goydos	22.2	P. Perez	30.0	V. Singh	42.9
2.	S. Ames	22.6	S. Maruyama	30.3	N. Lancaster	43.7
3.	K. Stadler	23.1	C. Pavin	31.1	D. J. Trahan	45.0
4.	O. Uresti	24.3	G. Day	32.7	D. Chopra	45.5
5.	B. Glasson	24.8	N. Green	33.1	R. Sabbatini	45.6
6.	T. Woods	25.0	J. Rose	34.2	M. Allen	45.7
7.	N. Price	25.2	S. Micheel	34.2	S. Ames	45.9
8.	T. Lehman	25.4	J. Brehaut	34.6	D. Love	46.1
9.	G. Kraft	25.6	J. Kelly	34.6	J. Gove	46.1
10.	J. Kaye	25.8	C. Riley	34.8	N. J. Brigman	46.3
	Average	34.1	Average	43.4	Average	55.8

We first look at the distance that the shots are off-line (left or right). Figure 9.5 plots the means (averages) of the distances off-line for various lengths of shots from the fairway. There is no surprise here: the longer the shot, the more off-line it is. From 170–180 yards out, the last point plotted shows that the average distance off-line is about 20 feet. As with figure 9.3, which shows average distances from the hole, an upward parabola nicely fits the distances beyond 50 yards.[6] That is, the general trend for the distance off-line is the same as for the total distance from the hole.

Figure 9.6 shows the standard deviations of the distance off-line from the same locations. Again, the character of the data changes at the 50-yard mark. There is actually a drop in standard deviation from the 50- to 60-yard range to the 60- to 70-yard range. While this could be an artifact of randomness, it may be the result of the pros being able to spin the ball better from distances beyond 50 yards, where the standard deviation is concave up.

Iron Byron: Approach Shot Statistics

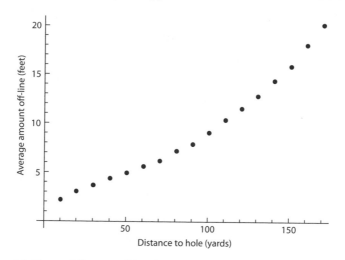

Figure 9.5 Means of distance off-line from 10–20 yards, …, 170–180 yards

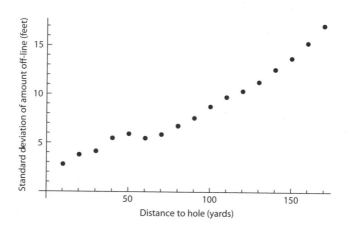

Figure 9.6 Standard deviations amount off-line from 10–20 yards, …, 170–180 yards

There is a surprise in figure 9.7, which shows the means of the amounts that the shots are short (negative values) or long (positive values). There are two choices here. By looking at absolute values, we can find the average amount that the distance is off. This would be analogous to what I did for figures 9.5 and 9.6. (There, I had no choice; I did not know whether a given shot was left or right.)

The other choice is to compute means of the actual values. In doing this, we are allowing a shot that is 10 feet too long to be cancelled by a shot that is 10 feet too short. The result is shown in figure 9.7. The pros, on the average, consistently come up short; the longer the shots, the shorter they fall. To keep it in perspective, coming up 6 feet short from 180 yards is not a disaster. However, I was not expecting to see the pattern in figure 9.7, where the means seem to oscillate.[7]

Common advice for nonprofessional golfers is to take more club than you think you need. If I'm 160 yards out, even though I usually hit 8-irons 155–160 yards, I should use a 7-iron. The logic is that most of my irons are less than pure, so I should count on "missing" enough with a 7-iron that it goes 160 yards instead of 170. The advice is probably good. However, I never plan on hitting weak shots and I would hate to "air-mail" a green because I actually made good contact.

Some of the "short syndrome" in figure 9.7 may be due to the same effect. On the rare occasions when a pro mishits an iron, the

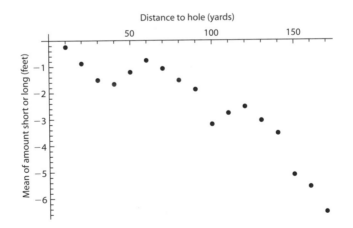

Figure 9.7 Means of amounts short (−) or long (+) from 10–20 yards, ..., 170–180 yards

shot is more likely to come up short than to go long. It is possible that there is more trouble behind a green than in front, in which case staying short may be a smart play. I can only speculate about the cause of the oscillations in figure 9.7, whereby the means go up and down as the distance increases. (There's probably a good explanation, but I don't have it.) That mystery remains unsolved.

TEN

The Big Dog
Driving Statistics

What other people may find in poetry or in art museums,
I find in the flight of a good drive.

—*Arnold Palmer*

A trip to Oakhurst Links in White Sulphur Springs, West Virginia, is a journey back in time to the 1880s. The setting is pastoral, both in the sense of a beautiful rural setting and in the sense of the herd of sheep that helps keep the links trimmed.[1] The 9-hole course at Oakhurst was restored to its original layout in 1994 by owner Lewis Keller and architect Bob Cupp. Dating back to 1884, Oakhurst was the first golf club in the United States. To honor that historic distinction, you play Oakhurst with hickory sticks and gutta-percha balls, imported from St. Andrews, Scotland. You form your own sand tees on the tee box, and the "stymie rule" is in effect.

I had the pleasure of playing Oakhurst in July 2008. Mr. Keller, at 85 years young, was a wonderfully upbeat, inquisitive, and gracious host. He was pleased to hear that this book would be about the mathematics of golf, as he had recently discussed similar ideas with Luke List, a Vanderbilt golfer who was runner-up in the 2004 U.S. Amateur championship at Winged Foot, Keller's old club. Mr. Keller has an enviably rich collection of friends and stories. As told in *Oakhurst: The Birth and Rebirth of America's First Golf Course* (coauthored by Vikki Keller, Lewis Keller's daughter), Sam Snead played an important role in the decision to restore Oakhurst Links, and he hit the first shot at the re-opening of the course.

The special club built for the occasion by Karsten Solheim did not perform well. The clubhead broke away from the shaft at impact and flew off along with the ball. Snead, never at a loss for words, immediately asked the shocked gallery, "Did either of 'em get on the green?"[2]

Newcomers to Oakhurst must adjust to the four-club "set" consisting of a driver, a niblick, a putter, and your choice of a cleek or a mashie. On the tee, there is the visual distraction of holding a "needlenose" driver with a head shaped more like a wooden banana than the modern metallic grapefruit. In spite of its relatively small size, the wooden driver head is very heavy, so that finding the right swing rhythm is a challenge.[3] On the green, you use a putter that looks disconcertingly similar to the driver. The putter is only slightly shorter but is much lighter and less lofted. The length and angle of the putter head are reminiscent of a hockey stick, which probably explains why a wristy putting stroke is effective.

Interestingly, the hickory shafts do not feel especially different from the shafts of today's clubs. The flexibility of hickory shafts is similar to the flexibility of modern graphite and soft steel shafts.[4] The gutta-percha balls are small and light and do not fly very far. In spite of the lack of dimples on the gutties, I can testify that monstrous slices are possible. Although the hickory sticks of the 1880s are recognizable as tools for golfing, the modern golf club represents a complete and extreme makeover from grip to clubhead. In this chapter, we explore how adept modern golfers are with their customized drivers.

CCs, MOIs, and CORs

The technological explosion has forced golf's governing bodies to take action. Drivers are now limited to a clubhead volume of 460 cc, and the coefficient of restitution cannot exceed 0.83. This has put club manufacturers in the awkward position of not being able to significantly change the most important characteristics of drivers, while needing to market a "revolutionary" new design.

The clubhead volume most directly affects the moment of inertia (MOI) of the club, although this is an imprecise statement, as the clubhead has different MOIs for different directions of rotation. For a given axis of rotation, the larger the MOI is, the more resistant the club is to rotation about that axis. The rotations that we want to minimize are the club twists caused by off-center hits. In effect, an increased MOI can enlarge the sweet spot of the clubhead. Due to the reduced twisting, even off-center hits can produce solid drives.

The coefficient of restitution (COR) is a measure of the "liveliness" of the driver. When clubheads were made of solid wood, the COR was determined by the type of wood used. With hollow metal clubheads, the driver face can be made thin and flexible, so that the golf ball receives a trampoline effect and springs off the clubface. The cap on the COR levels the playing field, in that most modern drivers are equally lively. The main question now is how well the characteristics of the club match a particular swing.[5]

One surprising aspect of driver construction is that small changes in the center of gravity of the clubhead can have large effects on launch angle and spin rate. This is one of many reasons to test-drive a new club and have a professional custom-fit you with an appropriate club. Many discussions about drivers focus on loft: Is it better to use a 9° driver or a 10°? As Tom Wishon argues in *The Search for the Perfect Club*, for most golfers this is the wrong question. A better choice for the typical golfer is between a 12° driver and a 14° driver. Each golfer has a typical swing speed and angle of attack, and for the ball speed and spin rate produced there is a precise angle that gives maximum distance. The important fact is that the ideal angle is different for different swing speeds. Wishon has shown that the slower your swing speed is, the more loft you need on your driver. An average golfer with a 90-mph swing speed is best served by a 13° or 14° driver.[6]

Tour Statistics

Since professional drivers are legislated to be similar in performance, it should not be surprising that driving statistics from the Tour changed little from 2004 to 2008. Figure 10.1a shows average driving distance, with measurements taken only on holes designated as "driving holes"—that is, holes on which most pros would actually hit driver. Figure 10.1b shows percentages of fairways hit over time. The vertical scales are very narrow, which exaggerates the differences in values, but the average driving distance has varied less than 1 yard from its 6-year average, and the percentage of fairways hit has varied less than 1 percentage point from its 6-year average.

Another driving statistic of interest is the longest drive of the year. Of the six years of my data sets, 2004 was the big year for

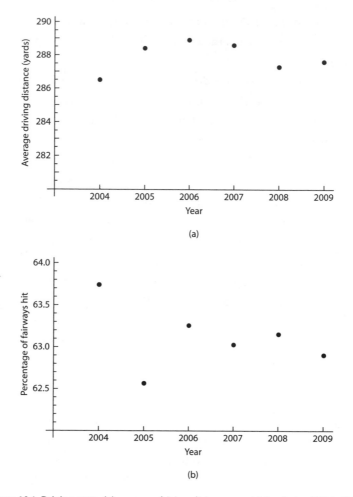

Figure 10.1 Driving stats: (a) average driving distance on driving holes, 2004–2009; (b) percentage of fairways hit, 2004–2009

long drives, including a whopping 476-yard launch by Davis Love III. The longest drives for the succeeding years were 442 (2005), 427 (2006), 437 (2007), 435 (2008), and 467 (2009) yards. In 2009, there were 32 recorded drives of over 400 yards. There are some big hitters out there.[7]

Par for a Two-shot Hole

The driving statistics vary with the length of the hole. The ShotLink data sets do not record which club is used, so we must infer whether a driver was used by the length of the shot hit. There is not a typical length for a par 4. Table 10.1 shows how frequently par 4s of different lengths were played in 2007. Most par 4s are between 380 and 490 yards in length. On the shorter holes, many of the pros will back off of the driver and hit an iron or hybrid off the tee. The length of the average drive depends on the length of the hole, as shown in figure 10.2. Once the length of the hole clears 350 yards, the length of the average drive climbs steadily from about 250 yards to 310 yards.

For some of the shorter hole lengths, the average drive distance is 10–20 yards longer for drives that ended up in the rough. This would seem to be backwards: if the ball gets into the rough, it should not

Table 10.1 Percentages of par-4 holes of different lengths, 2007

Length of hole (yd)	Percentage of all par-4 holes	Length of hole (yd)	Percentage of all par-4 holes
310	0.4	420	8.8
320	0.9	430	8.3
330	1.5	440	9.7
340	1.1	450	9.4
350	1.7	460	9.9
360	2.0	470	7.8
370	3.1	480	6.0
380	4.6	490	4.2
390	4.7	500	2.1
400	5.6	510	0.9
410	6.9	520	0.3

Note: Each length category includes holes with yardages greater than the previous data point up to and including the current data point.

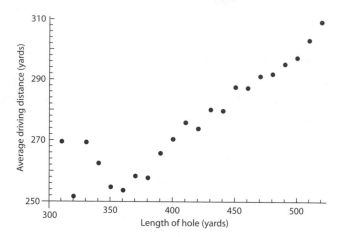

Figure 10.2 Average drive distance on par 4s of different lengths, 2007

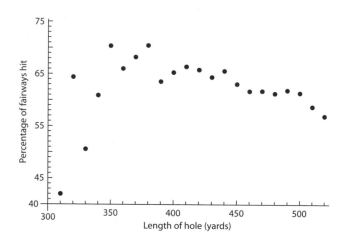

Figure 10.3 Percentage of fairways hit on par 4s of different lengths, 2007

roll as far. The apparent paradox resolves itself if you think of the causal relationship in the opposite direction. That is, the act of hitting the ball farther on short holes (for example, by hitting driver instead of a safer iron) increases the likelihood that the ball will find the rough. The percentage of fairways hit (see figure 10.3) also varies with distance, although not in a perfectly precise way. The fairway on a short hole can be hard to hit if it is very narrow, and the fairway on a long hole can be easy to hit if it is the size of Montana.

This is the background necessary to compute a "par" for holes of different lengths. For example, from 420 yards the average drive is 274 yards, leaving 146 yards to the pin. The average approach shot from 146 yards in the fairway is 25.3 feet from the hole (see chapter 9), producing an average of 1.93 putts (see chapter 7). Therefore, the average score on the hole when the fairway is hit is 3.93. From the rough, the average approach shot from 146 yards is 63.5 feet from the hole, producing an average of 2.2 putts and a score of 4.20. If 66% of the drives hit the fairway, then the average score on the hole is 66% of 3.93 and 34% of 4.20, or $0.66(3.93) + 0.34(4.20) \approx 4.02$. So the 420-yard hole really has a par of 4.02.

In the above example, missing the fairway cost our average golfer 0.27 strokes, slightly over a quarter of a stroke. The difference between the expected score from the fairway and the expected score from the rough is about 0.25 for most distances. To be sure, missing a fairway does hurt golfers' scores. However, it rarely pays to give up distance to increase accuracy, as you'll see in what follows.

The calculation of average score illustrated above can be done for other distances, as shown in figure 10.4. The calculated average score gets consistently higher as the length of the hole increases. If my only interest here was in showing that the pros' performances on par 4s depend on the length of the hole, I would simply compute their average scores on par 4s of different lengths. As shown in table 10.2, the match is quite close for most distances. There are consistent underestimates, starting at 420 yards and escalating when the hole length exceeds 480 yards, that are somewhat problematic. My predicted scores do not anticipate penalty strokes or major problems in the trees, and unanticipated disasters may explain the discrepancies in scores.

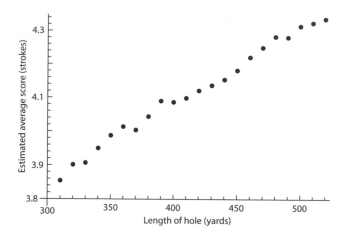

Figure 10.4 Estimated average score of par 4s of different lengths, 2007

The reason for computing the average scores is to set up the rating system that will be introduced in chapter 11. This rating system depends on our ability to predict an average score from different positions on the golf course.

Par 3 Statistics

To complete our analysis of tee shots, we examine par-3 tee shots next. One of the opportunities that the par-3 environment gives us is the chance to estimate how much landing in a bunker costs the pros. Not surprisingly, the distance to the hole after the tee shot depends on the length of the hole. Figure 10.5 illustrates the dependence. Figure 10.5 should remind you of figure 9.3 for approach shot distances from the fairway. Apparently, whatever sidehill, uphill, and scruffy lies the pros find in the fairway do not degrade their accuracy much from that enjoyed from the par-3 tee box.

Distance from the hole is only part of the story. As in real estate, location is also key. Of the 64,324 par 3s played in 2008, the pros hit the green 39,561 times (61.5%). The average score when they

Table 10.2 Predicted average scores and actual average scores for par-4 holes of various lengths, 2007

Length of hole (yd)	Predicted avg. score*	Actual avg. score
310	3.75	3.74
320	3.80	3.81
330	3.81	3.75
340	3.85	3.83
350	3.89	3.90
360	3.91	3.92
370	3.90	3.94
380	3.94	3.95
390	3.99	3.99
400	3.99	3.98
410	4.00	3.99
420	4.02	4.04
430	4.04	4.06
440	4.05	4.07
450	4.08	4.11
460	4.12	4.14
470	4.15	4.18
480	4.18	4.19
490	4.18	4.23
500	4.21	4.28
510	4.22	4.29
520	4.23	4.27

*Based on average drives, average approach shots, and average putts.

hit the green was 2.87. They landed in bunkers 6,411 times (10%), with an average score of 3.52. From tee shots that finished in the rough (6,841 of them, or 10.6%), the average score was 3.46. The pros actually scored better from lies classified as rough than from bunkers. From this evidence, the difference between hitting the green and dumping their tee shots into a bunker was more than half a stroke. The pros found water 983 times (1.5%), averaging 4.76 strokes. (This works out to a penalty stroke plus about a

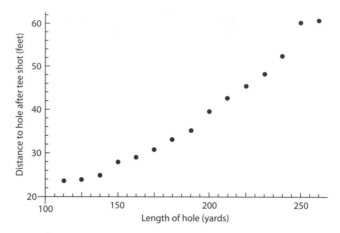

Figure 10.5 Average distance to hole after par-3 tee shots, 2004–2008

quarter of a stroke more than if they found the rough or a bunker.) From the fringe, the pros averaged 3.06 strokes, while from the fairway, they averaged 3.24 strokes. Corresponding statistics for par 4s show average scores of 3.01 from the green, 3.94 from the fairway, 4.01 from the intermediate rough, 4.27 from the primary rough, and 4.35 from bunkers. The average score for drives into fairway bunkers is worse than the average score from the primary rough. The average score for driving into greenside bunkers is 3.73, which is again more than half a stroke worse than being on the green. Bunkers cause more of a problem than I had realized.

Player Statistics

Turning to player statistics, we start with par-3 tee shots. Many of the issues that apply to approach shot statistics apply here. In particular, compiling average distances to the hole for all par 3s played can unfairly penalize players who play tournaments with longer par 3s. A more accurate measure is to use average distances better or worse than Tour average. Table 10.3 shows the top tens for 2007, 2008, and 2009.

Table 10.3 Top ten average par-3 tee-shot distances, compared to Tour averages, 2007–2009

Rank	2007	Shot distance (ft)	2008	Shot distance (ft)	2009	Shot distance (ft)
1.	T. Woods	6.56	T. Woods	5.97	J. Bohn	5.92
2.	E. Els	5.24	H. Slocum	5.16	R. Allenby	5.56
3.	J. Williamson	4.26	K. Perry	4.26	O. Wilson	5.01
4.	T. Clark	3.85	T. Byrum	4.03	K. Perry	4.92
5.	R. Moore	3.61	V. Singh	4.01	T. Pernice	4.56
6.	S. Verplank	3.50	C. Campbell	3.94	T. Woods	4.47
7.	R. Mediate	3.14	G. Ogilvy	3.90	S. O'Hair	4.23
8.	H. Slocum	3.11	R. Allenby	3.89	H. Stenson	4.09
9.	K. Perry	3.10	P. Casey	3.83	T. Lehman	4.07
10.	F. Funk	3.08	P. Mickelson	3.80	P. Sheehan	3.94

On average, in 2008 Tiger's par-3 tee shots finished 6 feet closer to the hole than the Tour average from the same distance. In 2007, Tiger distanced himself from the Tour by an even wider margin. As with irons from the fairway, however, Tiger's performance slipped in 2009. If you're wondering, Tiger's average par-3 tee shot in 2008 finished 32.2 feet from the hole, ranking 22nd on tour. While pure distance averages over all par 3s give little useful information, an interesting pattern emerges when the averages are broken down by distance. Table 10.4 shows top ten averages for par-3 tee shots over the 5-year span of 2004–2008 for the most common distances. From 180–210 yards, Tiger ranked 15th at 34.5 feet. From less than 150 yards, Tiger ranked 201st at 30.1 feet. Statistically, then, Tiger averaged about the same distance to the hole from all distances of 210 yards or less. Very odd.

Grip It and Rip It

Golf fans have always had a special fascination with long hitters. The special aura that goes with the long ball has been enhanced over the years by the characters who let it rip. The best example of the breed is John Daly.

John Daly became an instant hero when he dominated the 1991 PGA Championship. His victory was a true Cinderella story, and his good-old-boy charm magnifies his popularity.[8] But John Daly's legend is built on power—a huge swing that produces outrageous

Table 10.4 Top ten average par-3 tee-shot distances, 2004–2008

	Tee shot made from 150–180 yards		Tee shot made from 180–210 yards	
Rank	Player	Distance (ft)	Player	Distance (ft)
1.	T. Woods	26.1	K. Perry	32.9
2.	E. Els	26.6	H. Slocum	33.2
3.	S. Verplank	27.2	B. Glasson	33.2
4.	J. Rose	27.3	C. Campbell	33.5
5.	R. Gamez	27.4	F. Funk	33.5
6.	A. Scott	27.5	J. Sluman	33.8
7.	C. Campbell	27.7	E. Els	33.8
8.	F. Funk	27.8	R. Goosen	34.1
9.	R. Allenby	27.8	J. Durant	34.2
10.	K. Perry	28.0	J. Kelly	34.3

height and distance. Commenting on Daly's drives, Ian Baker-Finch said, "I don't go that far on my holidays," and Gay Brewer exclaimed, "Man, I can't even point that far."[9] Age and declining fitness have helped displace Daly as king of the long drives.

The driving distances given in table 10.5 are averages for the special driving holes on which the PGA predicts most players will use a driver. Five-year driving averages for both distance and accuracy are given in table 10.6.

For driving distance, there is little mystery in the fact that Watson and Holmes dominate. Quite elementary. In driving accuracy, Phil Mickelson was 180th at 58.5%, Tiger Woods was 191st at 57.5%, and the only two under 50% were David Duval (48.8%) and John Daly (49.0%). Top tens in driving accuracy for 2008 and 2007 are given in Appendix table A10.1. The top tens for distance and accuracy provide some perspective on the relative importance of these two measures. While accuracy is clearly important (we have seen that a missed fairway can raise the average score by a quarter

Table 10.5 Top ten driving distances, 2007–2009

Rank	2007	Driving distance (yd)	2008	Driving distance (yd)	2009	Driving distance (yd)
1.	B. Watson	316.4	B. Watson	315.1	R. Garrigus	312.0
2.	J. B. Holmes	312.9	R. Garrigus	311.0	B. Watson	311.4
3.	J. Daly	312.2	J. B. Holmes	310.3	D. Johnson	308.3
4.	R. Garrigus	311.0	D. Johnson	309.7	T. Ridings	307.4
5.	S. Gutschewski	307.5	B. Wetterich	304.1	G. Woodland	307.3
6.	T. Ridings	304.9	S. Allan	303.2	N. Watney	305.3
7.	H. Frazar	304.7	T. Ridings	302.9	K. Stanley	305.2
8.	B. Wetterich	304.1	N. Watney	302.9	R. McIlroy	305.2
9.	T. Woods	303.8	J. Daly	302.1	J. B. Holmes	304.6
10.	J. Gore	303.6	A. Scott	302.1	A. Cabrera	304.1
	Average	288.6	Average	287.3	Average	287.6

Table 10.6 Top ten driving distances and driving accuracy (fairways hit),
2004–2008

Rank		Driving distance (yd)		Driving accuracy (%)
1.	B. Watson	316.6	F. Funk	76.8
2.	J. B. Holmes	313.9	O. Browne	74.8
3.	R. Garrigus	310.3	J. Durant	74.4
4.	J. Daly	308.1	O. Uresti	74.3
5.	B. Wetterich	306.9	J. Coceres	74.1
6.	T. Woods	306.1	S. Verplank	73.8
7.	T. Ridings	302.8	L. Mize	73.0
8.	C. Hoffman	301.6	H. Slocum	72.9
9.	H. Frazar	301.5	B. Bryant	72.8
10.	J. Gore	301.3	G. Coles	72.6

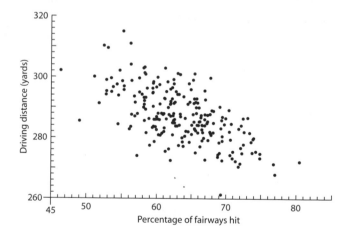

Figure 10.6 Accuracy (percentage of fairways hit) versus distance, 2008

of a stroke), you will find more major tournament winners on the
long-driving lists than on the most-accurate lists.

The scatter plot in figure 10.6 shows that most players excel
at distance or accuracy but not at both. The correlation between
distance and accuracy for 2008 averages is −0.62.

We will return to the issue of distance versus accuracy when we focus on rating players off the tee.

The Back Tee: The Right versus the Left

When players miss the fairway, do they consistently miss in particular places? The following quirky study was prompted by one of those odd coincidences of life. Shortly after reading about Ben Hogan reconstructing his swing to eliminate the left side of the golf course, I was told that Tiger Woods never misses to the left.[10] This struck me as an interesting thread to pursue. ShotLink records whether a drive has gone into the left rough or the right rough. It is then a simple matter to find the percentage of drives that miss on the left compared to the drives that miss on the right. There is no "good" value here—just some information about driving tendencies.

Of the top 230 golfers in my data set for 2008, 73 of them found the left rough more often than the right rough (table 10.7). David Duval, a right-hander, and Richard Green, a left-hander, lead the way with about two-thirds of their wayward drives to the left. Of Phil Mickelson's missed fairways, 53.8% of them were to the left. Only 32 of the 230 golfers had over 55% of their missed drives to the left. Three of these golfers (O'Hern, Stankowski, and Green) were in the top ten in 2007, when Bob Tway led the tour with 66.5% of his missed fairways to the left.

Fully 151 of the 230 golfers missed more often to the right than to the left. I do not know the percentage of golfers who play fades regularly, but either the right-handed golfers' fades are slipping out of the fairway or their hooks are not hooking enough. This would be the case with Rocco Mediate, who almost always hits a hook: in 2008, over 58% of Rocco's missed fairways were missed to the right. For 77 of the 230 golfers, more than 55% of missed fairways were to the right. As advertised, Tiger Woods missed to the right

Table 10.7 Top ten players missing fairways in the left rough, 2008

Rank		*Fairways missed to the left (%)**
1.	D. Duval	68.2
2.	R. Green	66.7
3.	O. Browne	64.9
4.	L. Janzen	64.5
5.	D. LaBelle	61.0
6.	L. Mattiace	60.7
7.	L. Mize	60.5
8.	P. Stankowski	60.2
9.	N. O'Hern	59.8
10.	J. J. Henry	59.2

*Of total fairways missed by each player.

more than twice as often as he missed to the left. (In 2007, 62.4% of his missed fairways were to the right.)

This might be an interesting statistic to study over time, to see if the skew to the right for the Tour golfers is consistent or just a one-year statistical oddity. If it is consistent, the cause could be swing patterns (as with Ben Hogan), course design (fairways sloping right, the hole being a dogleg left, or trouble to the left), or some other cause that I have not thought of. Regardless, many more players miss to the right than to the left.

One lesson you might learn from this is that, if you want to get close to a particular golfer at a tournament, join the gallery to the right of the fairway. Your favorite golfer might be headed your way.

ELEVEN

Tigermetrics
Player Rankings

He's a good player. Of course, when I was his age,
I won the U.S. Open and PGA Championship.

—Gene Sarazen, talking about Tiger Woods at age 20

Who was the best golfer in the world in the decade of the 2000s? In most sports, "Who is the best?" is a provocative question that can carry a bar conversation or an hour of talk radio. In golf, however, there was no doubt who was the best from 2004 to 2009. The rating system developed in this chapter will clarify how dominant Tiger was in the 2000s, while providing a way to answer future questions of who the best golfer is.*

Much of mathematics is devoted to determining a "best" answer, whether it is the best design of a golf club, the best investment strategy for a company, or the best golfer. What can distinguish mathematics from other methods of answering these questions is that mathematicians start with clearly defined objectives and logically progress to the answer. There can be disagreement as to whether the objectives are well chosen, but the objectives are the

---

*As you can undoubtedly tell, most of this book was written before Tiger's 2010 scandal hit the fan. I decided to leave most of it as written. As of now, it is questionable whether Tiger will regain his status as a clear number one. Chapter 11 may need to be interpreted as a historical study of an amazing dominance of the Tour by one player, a dominance that, due to the large number of outstanding golfers in an increasingly global sport, may never be seen again.

only sources of hidden agendas or biases that influence the answer. The task in this chapter, then, is to precisely define what we might mean by "best" golfer and to follow through to see how these objectives identify Tiger as the best in the 2000s. This, of course, is not exactly the way the scientific method is supposed to work. It is poor form to decide in advance what the answer is. Instead, the evidence is supposed to speak for itself, with the rational mathematician merely recording the answer.

A Smaller Problem

Many of the issues that arise when thinking about a player rating system also apply to determining the best putter. This narrower question has the advantage of not having an obvious answer. A raft of putting statistics was dumped on you in chapter 7, and there are many more rafts of similar statistics afloat. What is a reasonable way to combine all of these numbers into a single rating?

One way would be to take the most important of the statistics and base rankings on that. Putts per green in regulation predicts scoring the best, so we could use only that. The problem is that putts per green in regulation does depend on ball-striking ability. If you always aim at the pin and either miss the green or stick it tight, your putts per green in regulation score might be lower than a better putter who is constantly having to two-putt from 40 feet. An important objective in defining a putter rating is to try to avoid dependence on other skills, such as iron play and chipping.

The percentages of putts made from different distances are important statistics, and a player who rates highly from every distance must be a good putter. But how can this information be combined into a single number? We could add rankings like the PGA does for driving distance and accuracy. However, this assumes that all of the statistics that we add together have equal importance. Plus, in many cases, the difference between ranking 40th and ranking 80th is only one or two made putts. More important, along with listing the top ten, it would be nice to be able to say something about how much better the number-one player is than the number-ten player. The best way to guarantee that the final rating is meaningful is to insist that each statistic that enters into the calculation relates to the same, meaningful quantity. So, what is the most meaningful quantity in golf?

It's the Score, Stupid

When he was contending to be the best player in the world, Nick Faldo trundled down to Fort Worth to visit the great Ben Hogan. Over lunch, Faldo asked Hogan what he needed to do to win the U.S. Open.

"Shoot the lowest score, Nick," said Hogan.[1]

As with many of Hogan's rare utterances, this has a pure logical base that turns out to have profound consequences. The question that Bill James and others constantly pondered about baseball statistics was how they relate to winning. Obviously, a home run is better than a single, but how much better? The answer, they insisted, had to come in some fashion from an analysis of how games are won. Similarly, to do anything with the avalanche of golf statistics that is now available, we must have a good criterion for evaluating the statistics. Any reasonable statistic (cue the Hogan tape) must relate directly to the golfer's score.

For example, in 2007 Tiger Woods made 12 out of 27 (44%) of his putts between 7 and 8 feet. We saw in chapter 7 that this is below the Tour average. In fact, that percentage placed Tiger at 181 out of 230 golfers for putts between 7 and 8 feet. But what does this really mean? You could talk about Tiger not being that great a putter or about 27 putts not being enough from which to draw a conclusion[2] or any number of provocative discussion points. The bottom line, however, is simply that Tiger took more strokes than he should have when putting from 8 feet. The exact number of strokes can be measured. The Tour average from 8 feet was 53.1%. If Tiger had made that average percentage, he would have made $0.531 \times 27 \approx 14.3$ of his putts. Since Tiger made only 12 putts, he lost 2.3 strokes when putting from 8 feet.

We can do this from every distance. However, there is a crucial element missing from this line of logic. From longer distances, the important statistic is not so much percentage of putts made but total putts needed to hole out. A player who makes 20% of his 20-foot putts is not really ahead of the game if he three-putts 25% of the time from 20 feet.

Here is my system. For each distance up to 100 feet, compile the Tour's average number of putts taken when the first putt is of that distance. Then, for each player, take every green for the entire season and compare the number of putts on that hole to the Tour average. From 60 feet, the Tour average is 2.248 putts. If a player starts 60 feet away and two-putts, he gets credit for being 0.248

Table 11.1 Sample calculation of putts better/worse than average

Hole	Distance (ft)	Tour avg.	No. of putts	Value	Total
1	13	1.725	1	+.725	+.725
2	35	2.022	2	+.022	+.747
3	7	1.417	2	−.583	+.164
4	41	2.067	2	+.067	+.231
5	22	1.902	2	−.098	+.133

strokes better than the Tour average on that hole.[3] Add them all up, and you have the total number of strokes saved (or lost) by that player when putting. If you divide by the number of holes played and multiply by 18, you get the number of strokes saved or lost per round.

A sample calculation is given in table 11.1. This golfer starts by making a 13-foot putt on the 1st hole. I don't know or care whether it was for birdie, par, or bogey. A one-putt from 13 feet is 0.725 strokes better than the Tour average of 1.725 putts from that distance. On the next four holes, the golfer two-putts from a variety of distances. From 35 and 41 feet, a two-putt is better than the Tour average, and the difference is added to the golfer's running total. From 7 and 22 feet, a two-putt is worse than the Tour average, so the difference is subtracted from the golfer's total. Through five holes, the golfer is 0.133 putts better than average, for an average of 0.0266 putts better than average per hole. Multiplied by 18, the golfer's average is $0.0266 * 18 = 0.479$ strokes per round better than average. The number that would appear in table 11.2 for this golfer is 0.48.

Before looking at the ratings, think for a moment about what is involved and not involved in each player's putting rating. Most important, the rating does not depend directly on how the player gets to the green. Regardless of whether it was a green in regulation or a great shot or a conservative shot, the rating shows how the player putted on each hole compared to the Tour average.

Table 11.2 Putting efficiency ratings: top ten putts per round better than Tour average from the same distances, 2007–2009

	2007		2008		2009	
Rank	Player	Rating	Player	Rating	Player	Rating
1.	J. Parnevik	1.02	C. Pavin	1.01	L. Donald	1.06
2.	J. Delsing	1.00	B. Tway	0.96	T. Woods	0.99
3.	T. Clark	0.98	L. Donald	0.85	B. Gay	0.91
4.	T. Woods	0.75	T. Woods	0.84	J. Furyk	0.81
5.	Z. Johnson	0.72	A. Baddeley	0.64	B. Molder	0.80
6.	T. Ridings	0.65	D. Chopra	0.63	B. Snedeker	0.79
7.	F. Jacobson	0.65	M. Turnesa	0.61	M. Kuchar	0.79
8.	B. Snedeker	0.62	B. Crane	0.60	G. Chalmers	0.77
9.	R. Sabbatini	0.61	D. Hart	0.60	B. Faxon	0.76
10.	B. Estes	0.61	C. Collins	0.59	C. Couch	0.76

Therefore, the rating has isolated the putting performance of each player for the year. The main element that the rating does not control for is difficulty of greens. A player who plays only the most difficult courses (Tiger, perhaps) might be underrated here because short putts could be more difficult to make, and two-putts from long distances could be much more difficult.[4]

Table 11.2 shows the top ten in putting efficiency from 2007 to 2009. (Ratings for 2004 to 2006 are in Appendix A, table A11.1.) The ratings represent strokes per round. This means that in 2009 Luke Donald took one putt per round less than an average Tour putter would have taken from the distances that Donald faced. Although he never rated as the best putter for the year, Tiger Woods made the top ten in five out of six years. (He was ranked 17 in 2006.) Tiger is the only golfer in the top twenty in all six years. Brad Faxon, Ben Crane, and Stewart Cink made the top ten three out of six years. Table 11.3 shows putting efficiency ratings for the six-year period from 2004 through 2009. Brad Faxon, Aaron Baddeley, and Steve Stricker are often cited as the best putters on the Tour. In 2005, Stewart Cink (no. 8) fell just short of breaking Faxon's record of

362 straight holes without a three-putt when one of the slick greens at Pinehurst (North Carolina) ended his streak at 351 consecutive holes.

The logic behind the putter ratings can be followed back onto the fairway and then all the way back to the tee. For example, how could we isolate chipping ability? The ShotLink data might tell us that players X and Y were both in the fairway 90 feet from the pin on a given hole. Player X chipped up 11 feet from the hole, while player Y got a foot closer. One way to rate these shots is to say that player Y did 1 foot better, but our goal is to relate everything to scoring. Do you really have an advantage putting from 10 feet over putting from 11 feet? There is a very slight advantage, and comparing the Tour putting averages from these two distances shows us exactly how much of one.

There is a larger problem with using proximity of shot as the criterion. Suppose player X hits two chips to within 12 feet. Tour average from 12 feet is 32%, so there is no guarantee that he makes either putt. Player Y averages 12 feet on two chips, but one of them is 4 feet away, and the other is 20 feet away. The Tour average from 4 feet is over 91%, so he almost surely has one up-and-down. Since

Table 11.3 Top ten putts per round better than average, 2004–2009

Rank	Player	Rating
1.	Tiger Woods	0.794
2.	Brad Faxon	0.707
3.	Jesper Parnevik	0.668
4.	Greg Chalmers	0.661
5.	Ben Crane	0.654
6.	Brian Gay	0.642
7.	Aaron Baddeley	0.638
8.	Stewart Cink	0.634
9.	R. S. Johnson	0.617
10.	Steve Stricker	0.593

the Tour average from 20 feet is about 16%, he has a much better shot at two saves (about 14%) than does player X (about 10%).

The solution is to convert each shot to a score. For example, a chip to within 12 feet is assigned a score of 1.69, equal to the average number of putts from that distance. A chip to within 4 feet receives a score of 1.11, the average number of putts from 4 feet. So, you can see that the importance of chipping closer is not having a better view of the hole; it is having a better chance of making the putt. In this case, 4 feet is not just 8 feet better than 12 feet; it is 0.58 strokes better.

The ratings in table 11.4 are not necessarily for chipping. Shot-Link does not record the type of shot played (chip, pitch, flop, putt). The ratings average the results of playing shots from between 4 and 50 yards from the hole, from the fairway. For each player, each shot from this range is compared to the Tour average from that distance and lie. As discussed above, the comparison is not about feet from the hole; it is about how many putts would be needed from that distance. A positive value means better than average (fewer strokes needed), and a negative value means worse than average. All of a player's shot values are added together; this number is then divided by the total number of shots, producing the rating for that player.

Phil Mickelson is known for his imagination and spectacular recoveries. "Lefty" ranks second for the six year span of 2004 through 2009. In 2007, his average shot from the fairway from 4–50 yards out finished in a position that was 0.136 strokes better than the position for the average Tour player hitting from the same distance. By making eight such shots, Phil would save a full stroke compared to the average Tour player, assuming that they both are average putters. The six-year top ten would make a *very* strong leader board.

Table 11.5 shows the corresponding ratings for shots from the rough from 4–50 yards. These are the spectacular flop shots and delicate pitches from gnarly rough that the pros execute so well.

Table 11.4 Strokes per shot better than Tour average, 4–50 yards, from fairway

	2008		2009		2004–2009	
Rank	Player	Rating	Player	Rating	Player	Rating
1.	P. Harrington	0.123	B. Estes	0.139	S. Stricker	0.0877
2.	K. J. Choi	0.109	T. Woods	0.127	P. Mickelson	0.0834
3.	T. Woods	0.108	B. Gay	0.122	J. Furyk	0.0692
4.	S. Stricker	0.093	E. Els	0.114	E. Els	0.0672
5.	T. Immelman	0.085	C. Riley	0.113	T. Woods	0.0667
6.	M. Sim	0.083	O. Uresti	0.112	P. Harrington	0.0626
7.	S. Maruyama	0.080	J. Walker	0.111	S. Appleby	0.0613
8.	D. Forsman	0.078	J. Leonard	0.109	T. Immelman	0.0588
9.	T. Wilkinson	0.074	A. Kim	0.109	S. Garcia	0.0532
10.	T. Petrovic	0.071	T. Wilkinson	0.107	B. Estes	0.0528

Table 11.5 Strokes per shot better than Tour average, 4–50 yards, from rough

	2008		2009		2004–2009	
Rank	Player	Rating	Player	Rating	Player	Rating
1.	J. Kaye	0.116	D. Hart	0.102	C. Riley	0.0705
2.	J. Coceres	0.108	C. Franco	0.091	J.-M. Olazabal	0.0668
3.	T. Woods	0.107	P. Mickelson	0.070	S. Stricker	0.0622
4.	C. Riley	0.093	F. Jacobson	0.062	P. Azinger	0.0590
5.	B. Rumford	0.086	G. Day	0.060	B. Langer	0.0543
6.	B. Gay	0.080	D. Stiles	0.059	M. Brooks	0.0509
7.	V. Singh	0.079	F. Lickliter	0.059	P. Harrington	0.0504
8.	R. Goosen	0.074	M. Brooks	0.057	A. Oberholser	0.0486
9.	S. Stricker	0.073	C. Pavin	0.054	V. Singh	0.0459
10.	T. Immelman	0.070	S. Sterling	0.054	T. Woods	0.0456

Chris Riley has the best six-year average. His average shot from the rough from 4–50 yards out finished in a position that was 0.071 strokes better than the position for the average Tour shot from the same distance. Given 14 such shots, Riley would save a full stroke compared to the average Tour player.

Ironing It Out

The same procedure can be implemented for any range of distances. The results are not dramatically different from the tables of average distances recorded in chapter 10. The advantage of the new calculations is that they all use strokes as the unit of measurement. This allows me to use larger distance ranges. For example, suppose that Phil averaged 2 feet closer to the hole than Tiger from one distance, and Tiger averaged 2 feet closer to the hole than Phil from a different distance. Are they even? Not necessarily. An improvement from 6 feet away to 4 feet away gives a large putting advantage, while an improvement from 46 feet away to 44 feet away hardly makes a difference. By contrast, an improvement of 0.1 strokes is the same from any distance. For the rankings in table 11.6, I grouped all of the distances that might correspond to full swings with irons into a 50- to 200-yard category. The results from the fairway are given first, then the results from the rough.

At the risk of being redundant, I will give another example of what the above statistics measure. Let's say that Tiger is in the fairway 146 yards from the pin. ShotLink recorded 1,098 shots from this distance in 2007. On average, these 1,098 approach shots finished 25.3 feet from the hole. In turn, from 25–26 feet the Tour averaged 1.93 putts. Therefore, an average player would take 2.93 strokes from Tiger's position. If Tiger were to hit his shot 6 feet from the hole, we would use the Tour putting average from 6 feet of 1.32 putts and conclude that Tiger's iron was worth 2.32 strokes. Tiger's shot saved him $2.93 - 2.32 = 0.61$ strokes from the Tour average. By comparison, a conservative play to 35 feet away would be worse than average. Since the Tour averaged 2.02 putts from 35 feet, Tiger's iron was worth 3.02 strokes, and this shot cost him $2.92 - 3.02 = -0.10$ strokes. For the two shots, Tiger is ahead by $0.61 + (-0.10) = 0.51$ strokes, an average of $\frac{0.51}{2} = 0.255$ strokes

Table 11.6 Strokes per shot better than Tour average, 50–200 yards, from fairway and rough

Rank	2008 Player	Rating	2009 Player	Rating	2004–2009 Player	Rating
Fairway						
1.	T. Woods	0.071	T. Clark	0.067	T. Woods	0.0575
2.	D. Toms	0.047	S. Stricker	0.064	S. Verplank	0.0431
3.	S. Verplank	0.043	R. Sabbatini	0.050	S. Stricker	0.0424
4.	E. Els	0.043	F. Couples	0.049	T. Clark	0.0382
5.	R. Palmer	0.037	J. Leonard	0.045	J. Furyk	0.0350
6.	B. Gay	0.037	T. Woods	0.042	A. Cejka	0.0346
7.	M. Weir	0.036	A. Cejka	0.036	D. Toms	0.0345
8.	T. Clark	0.036	J. Bohn	0.034	J. Leonard	0.0319
9.	S. Stricker	0.036	J. Furyk	0.033	E. Els	0.0308
10.	J. Leonard	0.035	C. Campbell	0.033	B. Langer	0.0266
Rough						
1.	P. Harrington	0.064	R. Garrigus	0.093	J. Rose	0.0437
2.	J. Rose	0.063	S. Garcia	0.089	T. Woods	0.0431
3.	S. Verplank	0.060	A. Cabrera	0.062	S. Garcia	0.0427
4.	W. Austin	0.058	J. Kelly	0.061	D. Chopra	0.0384
5.	C. DiMarco	0.057	D. Lee	0.060	R. Sabbatini	0.0378
6.	C. Howell	0.053	J. Walker	0.058	V. Singh	0.0375
7.	R. S. Johnson	0.053	J. Byrd	0.055	W. Short	0.0343
8.	T. Clark	0.049	R. Mediate	0.053	K. Duke	0.0343
9.	G. Coles	0.048	M. Allen	0.050	P. Perez	0.0338
10.	A. Scott	0.045	R. Sabbatini	0.049	R. Garrigus	0.0330

per shot. The rating shows that for all of Tiger's fairway shots (2004 through 2009) from between 50 and 200 yards, his average was 0.0575 strokes better per shot than the Tour average. Given 17 of these shots, Tiger would save a full stroke over the average PGA golfer. For all of Tiger's shots from the rough (2004 through 2009) from 50 to 200 yards, his average was 0.0431 strokes better per shot than the Tour average.

Choosing the distance range 200–250 yards from the fairway, we get the results in table 11.7. Many of these shots would be second shots to par 5 greens.

One interesting consequence of tying all ratings to strokes is that it becomes meaningful to compare different categories. For example, is it more important to be strong inside of 50 yards or outside of 200 yards? While the ratings do not fully answer the question, some useful calculations can be made. For instance, in 2009 Bob Estes saved 0.139 strokes per shot inside of 50 yards. If he had had six such shots in a·round, he would have saved 0.834 strokes for the round. If Robert Garrigus had had six shots in the 200–250 yard range from the fairway in a 2009 round, he would have saved $6 \times 0.093 \approx 0.56$ strokes for the round. Estes would have saved more with his short game than Garrigus with his long game. This analysis, of course, is valid only if both players were hitting six of these shots in a round. However, it does give you a basis on which to make comparisons.

For the most part, the top ten in each category of iron shot have similar strokes-saved ratings. There is not an obvious bias toward

Table 11.7 Strokes per shot better than Tour average, 200–250 yards, from fairway

	2008		2009		2004–2009	
Rank	Player	Rating	Player	Rating	Player	Rating
1.	C. Villegas	0.086	R. Gamez	0.106	T. Woods	0.0850
2.	T. Clark	0.070	B. Jobe	0.095	E. Els	0.0648
3.	T. Woods	0.063	T. Gainey	0.078	D. Waldorf	0.0579
4.	J. Durant	0.061	C. Franco	0.075	S. Garcia	0.0550
5.	B. Watson	0.061	Z. Johnson	0.067	C. Franco	0.0536
6.	H. Mahan	0.059	T. Woods	0.066	B. Jobe	0.0491
7.	N. Flanagan	0.057	Q. Atwal	0.063	S. Ames	0.0453
8.	J. Bolli	0.055	M. Letzig	0.051	B. Watson	0.0440
9.	T. Immelman	0.053	G. McNeill	0.048	R. Allenby	0.0439
10.	A. Buckle	0.053	J. Byrd	0.048	P. Mickelson	0.0427

Table 11.8 Strokes per tee shot better than Tour average, par 3s only

	2008		2009		2004–2009	
Rank	Player	Rating	Player	Rating	Player	Rating
1.	H. Slocum	0.051	R. Allenby	0.062	E. Els	0.0462
2.	T. Woods	0.048	J. Bohn	0.058	T. Woods	0.0404
3.	H. Mahan	0.046	T. Woods	0.053	F. Funk	0.0391
4.	C. Campbell	0.045	T. Lehman	0.051	K. Perry	0.0387
5.	B. Tway	0.044	T. Pernice	0.049	R. Allenby	0.0356
6.	E. Els	0.042	K. Blanks	0.087	H. Slocum	0.0326
7.	P. Casey	0.042	B. Bryant	0.082	B. Glasson	0.0325
8.	V. Singh	0.039	H. Mahan	0.082	C. Campbell	0.0303
9.	G. Ogilvy	0.038	K. Perry	0.081	T. Clark	0.0302
10.	T. Clark	0.037	D. J. Trahan	0.076	J. Kelly	0.0302

either the short or the long game. As well, the numbers quoted above for a round are slightly less than but similar to the strokes saved in a round by the best putters. If Tim Clark had had 12 shots from the fairway between 50 and 200 yards in a round in 2009, he would have saved $12 \times 0.067 = 0.804$ strokes for the round, more than all but the top four putters in the putting efficiency ratings in 2009.

The same system can be applied to ratings for bunker shots. These are given in Appendix A, table A11.2. Table 11.8 shows the ratings for par 3 tee shots.

So far, we have analyzed putts and almost every shot with which the golfer would expect to hit the green. This leaves tee shots on par 4s and par 5s as the only common shots for which we have not accounted. These tee shots present a new challenge for relating the quality of the shot to strokes saved or lost.

Teeing Off on the Ratings

The concept for rating tee shots on par 4s and par 5s is the same as for the various shots discussed above. Take the position the golfer is in: in this case, it's the tee box for a hole of a certain length.

From the ShotLink data for the Tour, compute the average score that the average pro would make from that position. We did this computation for par 4s in chapter 10 for holes of different lengths. Then look at the drive that the golfer hits and compute the average score from where the ball finishes. Comparing the two averages, we get the number of strokes better than or worse than average for that drive. Add up all of the net strokes for all of the golfer's drives and divide by the number of drives.

To give one example, suppose that a golfer is on the tee for a par 4 hole of length 420 yards. The average score for that position is based on an average drive length of 274 yards and an average of 66% for drives landing in the fairway. As computed in chapter 10, this leads to an average score of 4.02. Suppose the golfer smokes a 300-yard drive in the fairway. From 120 yards out in the fairway, the average Tour approach shot lands 20.6 feet from the hole; from there, the pros average 1.87 putts. Following this drive, then, average shots will produce an average score of 3.87. The drive is therefore $4.02 - 3.87 = 0.15$ strokes better than average.

Table 11.9 shows top tens for driving on par 4s. If you glance down the lists, you may be struck by how they are dominated by the long hitters. However, the system is not intentionally biased. The ratings show us that, from a typical Dustin Johnson drive on a par 4 in 2009, the average player would score 0.058 strokes better than he would from his own average drive. If the long hitters dominate the list, that is good evidence that distance is more important than accuracy. This is well known to the top golfers, and it explains why the modern game is all about bombing away from the tee. It's fun, and it also pays off with better scores.

The other important fact to note about table 11.9 is that the stroke ratings are not very high. That is, nobody other than Dustin Johnson was gaining even a half stroke per round on the average golfer based on driving ability. This supports the findings in chapter 6 that driving stats do not correlate highly with scoring.

Table 11.9 Strokes per tee shot better than Tour average, par 4s only

	2008		2009		2004–2009	
Rank	Player	Rating	Player	Rating	Player	Rating
1.	V. Singh	0.039	D. Johnson	0.058	B. Weekley	0.0281
2.	C. Hoffman	0.035	C. Warren	0.040	J. Holmes	0.0267
3.	N. Watney	0.035	R. Fowler	0.040	C. Warren	0.0264
4.	J. Holmes	0.035	N. Watney	0.032	V. Singh	0.0248
5.	C. Warren	0.033	S. Allan	0.030	N. Watney	0.0221
6.	B. Weekley	0.031	J. Rollins	0.028	K. Perry	0.0214
7.	H. Frazar	0.030	M. Laird	0.028	L. Glover	0.0213
8.	D. Johnson	0.027	R. Fisher	0.027	G. Willis	0.0200
9.	J. Huston	0.027	A. Romero	0.027	B. Haas	0.0196
10.	A. Buckle	0.025	K. Perry	0.026	B. Watson	0.0189

The ratings for par 5 drives are computed in a similar way, except that they are based on a hole length of 520 yards. All of a player's drives on par 5s are applied against this standardized hole length and compared to the average Tour drive on par 5s. The results are given in Appendix A, table A11.3. and are similar to the driving stats for par 4s.

Overall Rankings

The original goal in this chapter was to identify the best player on the PGA Tour. What we have accomplished so far is to evaluate individual skills in terms of strokes better or worse than average. Using the common measurement unit of strokes, all we need to do to get an overall rating is some simple arithmetic. As an example, let's take Tiger Woods, who (surprise!) ranked number one overall for the 2009 season.

Tiger tees off on all 18 holes. On average, eleven of these are par 4s, four are par 3s, and three are par 5s. For 2009, his ranking of 0.0145 strokes better than average for par 4 tee shots is multiplied by 11, his rating of 0.0531 strokes better than average for par 3 tee shots is multiplied by 4, and his rating of 0.0932 strokes better than average for par 5 tee shots is multiplied by 3. Adding these values, Tiger is

$$11(0.0145) + 4(0.0531) + 3(0.0932) = 0.65$$

strokes better than average on tee shots. Fourteen of the holes are not par 3s; let's assume that on ten of these holes an approach shot between 50 and 200 yards from the fairway is required, and that the other four require a shot from the rough from 50–200 yards.[5] Let's also assume that six shots between 4 and 50 yards are required, with three from the fairway and three from the rough. These might be third shots on par 5s or recovery shots on par 4s. Tiger's 2009 ratings from 50–200 yards are 0.0419 strokes better than average from the fairway and 0.0109 strokes worse than average from the rough, while from 4–50 yards he rates 0.1271 strokes better than average from the fairway and 0.0401 strokes better than average from the rough. Add in one sand shot (rating 0.0597) and two shots on par 5s between 200 and 250 yards (rating 0.0527), and you have an overall approach shot rating of $10(0.0419) + 4(- 0.0109) + 3(0.1271) + 3(0.0401) + 0.0597 + 2(0.0527) = 1.08$ strokes better than average. His putting rating for a full round is 0.99 strokes

Table 11.10 Strokes per round better than Tour average, all shots considered

Rank	2004		2005		2006	
1.	T. Woods	2.19	T. Woods	2.21	T. Woods	2.21
2.	V. Singh	1.60	V. Singh	1.52	S. Stricker	1.74
3.	S. Cink	1.43	J. Furyk	1.42	J. Furyk	1.52
4.	R. Goosen	1.29	L. Donald	1.36	L. Donald	1.38
5.	A. Scott	1.24	E. Els	1.31	P. Mickelson	1.18
6.	G. Ogilvy	1.11	P. Mickelson	1.30	E. Els	1.13
7.	S. Maruyama	1.11	J.-M. Olazabal	1.24	S. Cink	1.10
8.	S. Ames	1.09	D. Toms	1.18	B. Gay	1.06
9.	J. Rose	1.03	S. Stricker	1.18	S. Ames	1.00
10.	P. Mickelson	1.02	S. Maruyama	1.11	T. Immelman	0.99

better than average. Adding this all together, for a round consisting of the shots described above, Tiger's expected score is 0.65 + 1.08 + 0.99 = 2.72 strokes better than average (rounded to two decimal places). Notice that in the one category in which Tiger rated worse than average, the rating is negative and points are deducted from his score.

This calculation is done for each of the players. The top ten for the 2004 through 2009 seasons are shown in table 11.10. Second tens for 2007 through 2009 are given in Appendix A, table A11.4.

Recall again that these ratings are based on ShotLink statistics, which only cover PGA Tour tournaments and exclude the four major tournaments and other tours. The rating is for one round of golf. So, the ratings say that in 2009 Tiger was 1.2 strokes *per round* better than anybody else on tour. In 2008, he was 1.29 strokes per round (more than 5 strokes for a tournament) better than anyone else. In each of the years from 2004 through 2009, Tiger rated more than 2 strokes better than average. He is the only golfer to reach the 2-stroke rating level. His dominance is spectacular. A brief analysis of the ratings for a small number of golfers is given in Appendix B. You will notice that Tiger is unusual in that in some years he rates as better than average in every category considered. Most golfers

2007		2008		2009	
T. Woods	2.72	T. Woods	2.58	T. Woods	2.72
S. Stricker	1.69	A. Kim	1.29	S. Stricker	1.52
R. Sabbatini	1.58	J. Leonard	1.18	T. Clark	1.42
P. Mickelson	1.45	P. Mickelson	1.17	J. Furyk	1.27
E. Els	1.39	V. Singh	1.14	Z. Johnson	1.23
S. Garcia	1.30	B. Tway	1.11	D. Johnson	1.17
T. Clark	1.20	P. Harrington	1.04	A. Kim	1.15
A. Scott	1.17	K. J. Choi	1.02	G. Ogilvy	1.13
F. Jacobson	1.09	S. Stricker	1.02	P. Harrington	1.12
K. J. Choi	1.08	S. Appleby	0.95	L. Donald	1.11

have definite strengths and weaknesses, at least according to the numbers.

We humans have a love-hate relationship with numbers. While many of us can spend hours absorbed in the types of statistics detailed in this chapter (and may find ourselves in full command of Tiger's career statistics while stumbling when asked how many years we have been married), we also have a distrust of the avalanche of numbers that threaten to take away the mystique of the game. When a crucial putt slides in the side door, we tend to talk more about the golfer's strong will and determination and less about luck. Mathematicians as a group fall squarely on both sides of this issue. The profession is devoted to looking for patterns, many of which are numerical in nature. As descendants of Pythagoras and Galileo, we think that the world can be quantified. At the same time, mathematicians have more experience than most with various forms of "numerology" in which numerical coincidences are falsely given great significance. An essential part of a good mathematical analysis is an evaluation of the likelihood of errors in the analysis.

So, I am not going to point to Vijay Singh's 2008 putting rating of −0.41 and say that he is a poor putter. For one thing, I have a healthy respect for two of Vijay's other statistics: at 6 foot 3 and 208 pounds, he is a lot bigger than I am. More importantly, the putter rating is only for 2008 and is relative to the other players on the PGA Tour. This is a high standard. For his part, Singh has the right attitude. After winning the 2008 FedEx Cup, he said in an interview that, "I believed in myself that I'm the best putter."[6]

There are two main criticisms I have of the numbers presented here. One is the lack of consideration of course difficulty. Putting statistics and all of the other stats we've looked at can be affected by course setups, weather conditions, and other factors. Presumably, this means that golfers who tend to play in the toughest events (Tiger, Phil, most of the non-PGA–Tour players) are underrated in this rating system. The second criticism involves location. The system I have used so far is fairly coarse in that I have not taken into account finely detailed aspects of position. For example, when a shot finishes 30 feet from the hole, I use that distance in the calculation but don't factor in whether the ball is actually on the green. A better system would account for whether the ball is on the green or in a bunker. This type of fine-tuning is high on my to-do list.

The end-of-chapter discussion that follows is a first look at possible corrections for course difficulty. It is essentially a case study to assess how important such corrections might be.

The Back Tee: Poa Putting Conditions

The basic question is whether there is evidence that some courses play harder than other courses. This is not as silly a question as you might think. Scores are much higher in some tournaments than in others. Isn't that evidence enough? Actually, no. My real question is whether the course difficulty would affect the statistics that I've used in my rating system. If a course is just brutally long, then the pros will be hitting approach shots from outrageous distances and will score poorly. However, my ratings are based on comparing shots of the same distance. This will not hurt anybody's rating. Still, it would be naive to expect that all courses putt the same. Some greens are trickier to read, some greens are contoured more, and some greens are in better shape than others. Is there a way to measure the effects of difficult greens and adjust for them?

As a first attempt at assessing how strong the effects are, I computed a putting rating for each tournament in 2007 and 2008. That is, I treated each tournament as a golfer and ran all of the putting outcomes through the putting efficiency formula. The results were more dramatic than I was expecting. In 2008, all four tournaments in California rated as harder to putt than average. Six of the seven Florida tournaments rated as easier to putt than average. Three of the four Texas tournaments rated as easier to putt than average. Two-thirds of the tournaments showed the same effect (harder or easier) in 2007 and 2008. My conclusion is that there are some stable, measurable putting effects that should be incorporated into a good rating system for putters. In fact, a group of MIT professors has done just that. Douglas Fearing, Jason Acimovic, and Stephen Graves created a statistic called "putts gained per round" using ShotLink data.[7] Their approach is similar to what I have done, but they corrected for course difficulty.

My computations indicate that Pebble Beach is the most difficult course for putting, rating 1.0 and 0.7 strokes per round harder than average in 2007 and 2008, respectively. In both years, Riviera rated 0.4 strokes per round harder to putt than average. Torrey Pines rated 0.3 and 0.5 strokes per round harder than average in 2007 and 2008, respectively. Note that all three of these courses are in California. I was curious about what was common to the courses that rated as difficult. Any course can be made impossible to putt; we have all suffered through "greenskeeper's revenge" days where the pins are all placed in the worst locations imaginable. But the fact that the difficulty ratings seemed to depend on geography made me suspicious that the type of grass might have something to do with it. Although I am not at all knowledgeable about how different strains of grass might affect putting, it is interesting that Pebble Beach, Riviera, and Torrey Pines list poa annua as a primary grass used on the greens. In fact, five of the six courses that list poa annua for greens rate harder than average in both 2007 and 2008.

One hard course that does not feature poa annua greens is the Plantation Course at Kapalua, Hawaii. The rating was 1.2 strokes harder than average in 2007 and 0.3 strokes harder than average in 2008. The Plantation Course is often buffeted by high winds, which can make putting a nightmare. Without having weather reports, I would suspect that wind played a role in the very large difference in the difficulty ratings for the two years. Also, the season-opening tournament played at Kapalua has a small field, which could lead to large deviations in statistics.

The easiest course in the ratings was the TPC Deere Run, rating 0.4 and 0.5 strokes per round easier than average in 2008 and 2007, respectively. The largest single-tournament effect on the easy side was the 2007 Colonial, at 0.7 strokes per round easier than average. The 2008 rating dropped to 0.2 strokes easier than average. Both Deere Run and Colonial have bent grass greens. Of the 23 courses listed as having bent greens in 2008, 15 of them rated as easier than average to putt.

Further studies into the effects of physical course characteristics on putting and all other shots are needed. The rating system presented here is a skeleton that will be fleshed out as the analysis of golf statistics progresses.

TWELVE

More Rating Systems
and Tiger Tales

When it comes to the game of life, I figure
I've played the whole course.

—Lee Trevino

Lee Trevino used to characterize the difference between himself and Jack Nicklaus by describing typical par 5 birdies for each. Nicklaus would launch a majestic drive into the fairway, hit a towering long iron onto the green, and narrowly miss his eagle putt. Trevino would slice a drive into the trees, punch an iron across the fairway into the left rough, gouge a wedge onto the green, and curl in a 10-footer. The game, Trevino would conclude with a grin, can be played in many different ways.

In chapter 11, I presented an intricate rating system and concluded that Tiger Woods was the best player on the PGA Tour in 2009 (and 2008, and 2007, and . . .). There are numerous other routes to the same conclusion. Some are majestic and some involve a lot of scrambling. A small sampling follows.

There are some obvious flaws in the rating system described in chapter 11. One flaw is that I have not given it a clever name. For reference, I'll call it the Total Strokes rating system. The TS ratings are influenced by how I divided a typical round into 10 approach shots from the fairway (50–200 yards), 4 approach shots from the rough, and so on. These numbers were chosen based on Tour averages for the number of shots hit of each type, but other choices could be made. More troublesome is the dependence of the TS

ratings on ShotLink data, which does not include the four majors. This problem is addressed in a different rating system, which I'll present next.

The only data I have available on the major tournaments are the scores posted by the players. To develop a rating system based only on scores, I modified a system that I have used for years for college football.[1] The first step is to recognize some inadequacies in the standard measures of golfing success. The Vardon Trophy is given annually to the golfer with the lowest scoring average. It does make some sense to honor the golfer with the lowest scoring average as the best golfer, but this statistic does not take into account the difficulty of the courses played. The top golfers play a high percentage of their rounds in big tournaments on demanding courses, where even par is often in contention for the championships. At some other events, even par may not make the cut.

The public tends to pay less attention to the scoring average and more attention to rankings such as money earned, World Golf Rankings, and FedEx Cup standings. All of these measures give extra weight to important tournaments and place extreme emphasis on winning. Money and points for first place can be double what they are for second place. If someone snakes in a monstrous putt to win a playoff, is that really twice as good as the unlucky second-place performance?

This is an especially interesting question as it relates to consistent and inconsistent golfers. Think about how you would rank the following two golfers. Player A makes the cut every week, usually finishing between 25th and 50th. Player B misses the cut over half the time but gets hot for a month or two and posts several top ten finishes with one win per year. Player B's wins attract more publicity, money, and points, but which golfer is really better? Both the TS ratings and the ratings that follow reward the strong play that results in a tournament win but award no extra credit for the win itself.

In its most basic form, the rating system I present here is very simple. Suppose that every time golfers A and B play in the same tournament, A beats B by 2 strokes. Based on this evidence, rating player A as 2 strokes better than player B makes sense. Following through on this logic, think of every tournament that Tiger plays in as a competition between Tiger and each of the other golfers. If Tiger shoots 283 and Phil shoots 286, then Tiger beats Phil by 3 points. If Trevor shoots 280, then Tiger loses to Trevor by 3. For every tournament and every golfer, compile all of the points won and points lost in all of these head-to-head comparisons.

Suppose that on average Tiger wins by 14 points per match-up. Then Tiger's rating should be 14 points higher than the average of his opponents' ratings. If, on average, Phil wins by 8 points per match-up, then Phil's rating should be 8 points higher than the average of his opponents' ratings. There is a mathematical procedure for determining all of the golfers' ratings so that each rating matches the player's actual results. Some details are given at the end of this chapter. I applied this system to tournament scores for 222 golfers in 38 tournaments from 2007, including the four majors.

Table 12.1 gives the top ten strokes ratings for 2007. The units here are strokes per tournament. Tiger is rated 14.2 shots above average for a four-round tournament. More remarkably, he rates more than 4 shots per tournament better than second-place Ernie Els. In the 2007 TS ratings Tiger was about 4.1 strokes better than second-place Steve Stricker for a four-round tournament and 10.8 strokes better than average for a full tournament. The extra dominance shown here may be the result of a strong record in majors (one win, two seconds, and a twelfth). It may also indicate that, on top of his other skills, Tiger manages his rounds so well that his total is greater than the sum of its parts.

A critical difference between these ratings and, for example, the World Golf Ratings is that if two golfers tied for the lead after four rounds, this system would rank them as tied regardless of which

Table 12.1 Top ten strokes ratings, 2007

Rank	Player	Rating*
1.	Tiger Woods	14.2
2.	Ernie Els	9.8
3.	Phil Mickelson	8.1
4.	Jim Furyk	7.6
5.	Steve Stricker	7.4
6.	K. J. Choi	7.3
7.	Sergio Garcia	6.8
8.	Vijay Singh	6.8
9.	Arron Oberholser	6.7
10.	Adam Scott	6.4

*Strokes better than Tour average for a four-round tournament

Table 12.2 Win/loss ratings, 2007

Rank	Player	Rating
1.	Tiger Woods	1.00
2.	Ernie Els	0.73
3.	Justin Rose	0.63
4.	Jim Furyk	0.61
5.	Sergio Garcia	0.60
6.	Vijay Singh	0.57
7.	K. J. Choi	0.57
8.	Steve Stricker	0.56
9.	Arron Oberholser	0.55
10.	Phil Mickelson	0.55

one won the playoff. This seems less than ideal, since winning is important and can be a measure of the player's ability to perform under extreme pressure. To give winning a more prominent role, the same system can be implemented using only wins and losses. For example, instead of giving Tiger 3 points for beating Phil by 3 strokes, give him 1 point for winning. When the points are counted this way, the top ten list changes, as shown in table 12.2.

This version of the ratings is harder to interpret. What meaning can be attached to Tiger being 0.27 wins better than Ernie Els? There is not much to say, except that it looks like Tiger is *way* better than anyone else. What I find most interesting about this approach is the actual data. For the tournaments involved, Tiger finished 2007 with a won-lost record of 1255-85. That computes to a 94% winning percentage, compiled in the majors and other top tournaments. Second place in terms of winning percentage was Ernie Els, well behind at 80%.

An important aspect of either version of these ratings is that the quality of the opponents (called "strength of schedule" in other sports) is factored into the system. In the strokes version of the ratings, if you beat an opponent by 2 strokes, your rating will be 2 points higher than your opponent's rating. The better your opponent's rating is, the higher your rating will be (since you get your rating by adding 2). The same phenomenon holds for the won-lost version. Beating a good player boosts your rating by more than beating a mediocre player.

When you have two rating systems, the inevitable question is which one is better? The tests that I ran are inconclusive on this question, other than to say that a combination of the two outperformed either one individually. The combination that did well is essentially an average of the two. Looking at the top tens, you can see that multiplying the won-lost ratings by 14 comes close to matching the values in the stroke ratings. The combination rating is 14 times the won-lost ratings plus the stroke ratings divided by 2. If you average the two, the ratings can still be interpreted as number of strokes above average for a four-round tournament.

The claim that the combination rating system outperforms the individual ratings needs to be justified. In fact, the more important claim is that it outperformed the FedEx ranking system for

Table 12.3 Combination ratings, 2007

Rank	Player	Rating*
1.	Tiger Woods	14.1
2.	Ernie Els	10.0
3.	Jim Furyk	8.0
4.	Phil Mickelson	7.9
5.	Steve Stricker	7.6
6.	K. J. Choi	7.6
7.	Sergio Garcia	7.6
8.	Justin Rose	7.6
9.	Vijay Singh	7.4
10.	Arron Oberholser	7.2

*Strokes better than Tour average for a four-round tournament

predicting the results for the 2007 FedEx playoffs. In particular, I computed the strokes and wins/losses ratings using only the events preceding the FedEx playoffs and used them to predict the outcomes of the playoff events. For example, Tiger was rated above Phil, so the ratings predict that Tiger beats Phil. This prediction turned out to be correct in Atlanta but incorrect in Boston. For comparison, I also used the FedEx standings going into the playoffs to predict the outcomes of the playoff events. The combination ratings system had the best prediction record. To be fair, the combination system got 59% of its predictions right versus 58% for the FedEx system and for both of the individual ratings systems. However, over the entire playoffs, a 1% difference translates into nearly 400 more correct predictions.

OMG, It's the OWGR

The topic of this section is the Official World Golf Rankings, a ranking system that is recognized and endorsed as the official golf ranking system by nearly all of the important golf organizations and tours. OWGR is a points system whereby a player's performance in a tournament earns points based on that player's placement in the tournament and on the importance of the tournament. Points earned in a tournament are kept for two years, although after 13 weeks the value of points from a tournament begins slowly decreasing, reaching 0 at the two-year mark. A player's rating equals his average number of points per tournament played.

For major tournaments, 100 points are earned for first place, 60 points for second place, 40 points for third place, 30 points for fourth place, and so on. Even though only one stroke over four rounds may separate each place, the point differential between places is substantial.[2] As with the money awarded, each place earns about 60% of what the next highest place earns. The Tour Championship winner receives 80 points, with points for second place, third place, and so on decreasing by similar percentages as for the

Table 12.4 Top ten point totals in Official World Golf Rankings, end of 2007

Rank	Player	Rating
1.	Tiger Woods	19.6
2.	Phil Mickelson	8.7
3.	Jim Furyk	6.6
4.	Ernie Els	6.5
5.	Steve Stricker	6.4
6.	Justin Rose	6.0
7.	Adam Scott	5.8
8.	Padraig Harrington	5.6
9.	K. J. Choi	5.1
10.	Vijay Singh	5.1

major championships. Other tournaments have maximum point values reflecting their relative importance world-wide.

The most important characteristic of the OWGR is that it includes tournaments from all over the world. The OWGR point system is relatively easy to understand and provides a useful starting point for conversations about golfer rankings. The top ten in OWGR points at the end of the 2007 season are shown in Table 12.4. The OWGR are surprisingly similar to the combination system rankings. In particular, the top five are identical except for the Phil and Ernie rankings. More information about the OWGR can be found at the OWGR website.

Berry/Larkey Ratings

One question that has not yet been addressed is that of ranking golfers over time. The rating systems that I have presented can be applied over any time frame for which the data are available. The TS ratings are time-limited because ShotLink data is a 21st-century phenomenon, while the combination rating system can be applied over very long time frames. It is interesting to think about how that might work.

Suppose that I want to compare Tiger Woods to Gene Sarazen. Given tournament data going back to the 1930s, I would have head-to-head match-ups between Woods and Jack Nicklaus, Nicklaus and Sam Snead, and Snead and Sarazen. The combination rating system could then rank everybody who competed in any of these tournaments. I did not try this, because there is reason to believe that these ratings would not be fair.

The 2000 PGA Championship had one of the most exciting finishes ever, with Tiger defeating Bob May in a playoff after both made dramatic putts on the 72nd hole. In the first two rounds of that tournament, Tiger and Jack Nicklaus were paired together, as Jack played his last PGA Championship. Their exchange on the 18th green of the second round was out of a Hollywood movie, as Jack acknowledged a huge ovation and with a gesture passed the mantle of greatness to Tiger. The point of this story is that while this gives us two Woods-Nicklaus head-to-head tournament rounds, they occurred when Tiger was in his prime and Jack was near retirement. Similarly, the age difference between Nicklaus and Snead means that most of their shared tournaments caught Jack in or approaching his prime and Snead trying new putting strokes to try to stay competitive. The comparison is not fair.

In their 1999 article "Bridging Different Eras in Sports," Berry et al. present a statistical model for estimating the effects of aging on ability in golf, baseball, and hockey. Their findings are fascinating in many ways. For golf, they find that the average trend for a golfer is somewhat like the simpler graph shown in figure 12.1.[3] For most golfers, the peak scoring years are between age 30 and 40, with a learning curve for younger golfers at around age 20 and a decline for older golfers at around age 50 which results in an average of 2 strokes per round higher than at the peak of their careers. If an aging function could be computed for each golfer, then comparisons over time could become meaningful. For example, in 1960 Jack Nicklaus and Ben Hogan played the last rounds of the U.S. Open together. Nicklaus was 20 years old and had not yet mastered the science of course management. Hogan was 47 years old and was struggling with his nerves on the greens. If their aging patterns were normal (they were not), we could take their scores from that tournament and subtract 2 strokes per round, which would provide a fair comparison to golfers in the field who were in their prime.

Of course, not everyone has the same aging pattern. Ben Hogan did not hit his stride until later in life, while Ben Crenshaw peaked at a younger age. Berry et al. allowed the aging pattern for a given

Figure 12.1 An aging function for golfers

golfer to change in several ways. Each player has a maturation age m_A and a declining age m_D. In between m_A and m_D, the golfer ages according to a more sophisticated version of figure 12.1. For ages younger than m_A, the value of the generic aging function is multiplied by a constant c_1, and for ages older than m_D, the generic aging function is multiplied by a constant c_2. Ben Crenshaw, who matured early, was only about 1 stroke above peak level at age 20; for him, $c_1 = \frac{1}{2}$. Ben Hogan aged well and was only 0.5 strokes above peak level at age 50; for him, $c_2 = \frac{1}{4}$.

The assumption is that a player's score in a given round is given by score = peak ability + round adjustment + aging function + error. The round adjustment accounts for the difficulty of the course on that day, as measured by the average score of the field. Berry et al. took scores from the four majors from 1935 to 1997 and statistically estimated the values of the variables that best fit the data. The phrase "best fit" means that the "error" terms in the formula are as small as possible.

The values found for the peak abilities of each golfer can then be used to give an all-time ranking of golfers. Most discussions of whether or not an athlete belongs in the Hall of Fame include evaluations of longevity as well as peak ability. Since the standard aging model shows golfers maintaining peak ability for 10 years or more, peak ability in this case does include some information about strong performances over time.

The system is designed to eliminate bias from one generation to the next, whether it be from redesigned clubs or courses of varying difficulty. However, to my mind the list of top 20 players given in table 12.5 is heavily skewed to modern golfers. So, let the debate begin. Are modern golfers better than their predecessors? Remember that these ratings were published in 1999 using data through 1997. At this stage, Tiger Woods had won exactly one major. Berry et al. note Tiger's young age and say that if he were to improve as much as the generic aging function predicts, he would

Table 12.5 Berry/Larkey ratings for peak
ability in majors, 1935–1997

Rank	Player	Rating
1.	Jack Nicklaus	70.42
2.	Tom Watson	70.82
3.	Ben Hogan	71.12
4.	Nick Faldo	71.19
5.	Arnold Palmer	71.33
6.	Greg Norman	71.39
7.	Justin Leonard	71.40
8.	Ernie Els	71.45
9.	Gary Player	71.45
10.	Fred Couples	71.50
11.	Hale Irwin	71.56
12.	Calvin Peete	71.56
13.	Julius Boros	71.62
14.	Raymond Floyd	71.63
15.	Lee Trevino	71.63
16.	Sam Snead	71.64
17.	Jose-Maria Olazabal	71.69
18.	Tom Kite	71.71
19.	Ben Crenshaw	71.74
20.	Tiger Woods	71.77

reach a peak score of 68.6 and rate as far and away the greatest ever. This is what has happened.[4]

Maximum Sigma

An estimation of peak ability is one way to attempt to answer the question of who is the greatest of all time. John Zumerchik approaches the problem in a different way.[5] He attempts to identify the greatest performance in a single tournament using standard deviation as his basic tool. Recall that if the distribution is normal (golf tournament scores are generally close to normal), the chance of getting a score 1 standard deviation below the mean is about

16%, 2 standard deviations, about 2.3%, and so on. This gives us a way to measure how good a score is. Compute the mean and standard deviations of scores for a tournament. The more standard deviations below the mean a score is, the more unlikely the score is and the more impressive the result. Zumerchik did this for a number of Masters tournaments and U.S. Opens through 2000.

Zumerchik's analysis found only five tournaments in which the winning score was 3 or more standard deviations below the mean. Three occurred in the Masters, all by players breaking the scoring record. In 1965 Jack Nicklaus shot 271, which was 3.56 standard deviations below the mean. Raymond Floyd matched the 271 in 1976, finishing 3.27 standard deviations below the mean. Tiger Woods shot 270 in 1997, winning by 12 strokes with a score 3.26 standard deviations below the mean. In the 1953 U.S. Open, Ben Hogan's 283 was 3.32 standard deviations below the mean.

Hogan's 1953 season is one of the candidates for best year by a golfer. He won the Masters, U.S. Open, and British Open and did not compete in the PGA Championship. Three other seasons offer candidates for best year ever, including Bob Jones's Grand Slam year of 1930 and Byron Nelson's 1945 season with 18 tournament victories, including 11 straight. If you can name the other (fairly recent) remarkable season, you can probably answer the question that is still on the table—identifying the greatest single tournament ever.

In 2000, Tiger Woods won three majors, breaking the scoring record in each. Does this mean that his performances were the best ever? If the courses were just playing especially easy that year, Tiger's records may not be that noteworthy. Tiger's score of 272 in the 2000 U.S. Open at Pebble Beach won by 15 strokes. Moreover, it was an unprecedented, unapproached, and nearly unimaginable 4.75 standard deviations below the mean. Given the scores the field posted at Pebble Beach that year, the probability that anyone would shoot 12 under par is about 0.000001. This is almost literally one in a million!

The Tiger Effect

You would not expect that a job market paper by a graduate student in the Department of Agricultural and Resource Economics would be of much interest to golf fans. Nevertheless, Jennifer Brown's 2007 paper "Quitters Never Win: The (Adverse) Incentive Effects of Competing with Superstars" created a national media stir with its conclusion that the PGA's best players score higher when Tiger Woods is in the field than they do when he takes the week off.[6]

Brown's intent was to explore an economic principle, that incentive bonuses can backfire. The naive thinking is that offering a bonus for the most sales will spur all salespeople to perform better than usual as they vie for the bonus. Brown wondered what would happen if the winner of the bonus was not in doubt due to the presence of a "superstar" salesperson. Having no realistic chance to earn the bonus, a good (but not great) salesperson not only would have little motivation to work extra hours but might in fact succumb to negative psychological impulses and actually sell less than usual. Brown had the inspiration to use tournament golf data to test this hypothesis. If Tiger is not playing, many good golfers can realistically expect to have a chance to win. They will prepare for the tournament with enthusiasm and be willing to grind out every stroke. If a dominating Tiger is in the field, perhaps they "give up" and play conservatively for a decent pay check.

At first glance, the "Tiger effect" appears to be dramatic. Limiting the data to the top 141 (exempt) Tour players, in 2006 the stroke average in tournaments with Tiger in the field was 2.73 under par, improving to 4.16 strokes under par when Tiger was not in the tournament. Be careful here! It would be easy to think that Tiger's presence cost his opponents a stroke-and-a-half, but this would ignore the fact that Tiger does not play on a random assortment of courses. The courses for Tiger events are typically harder than those for non-Tiger events. If the courses are, say, a stroke-and-a-half harder, then Tiger himself is not creating any effect at all.

This is one of the reasons that Brown's problem is challenging. There *is* a difference in performance when Tiger plays compared to when Tiger sits, but what is the real cause of it? Is it the difficulty of the course? Out-of-control fans in Tiger's gallery? To try to remove the course difficulty factor, Brown looked at tournaments that Tiger usually plays. If he does not play a Buick Open one year and the scores are good, then plays it the following year and the scores are bad, maybe the root cause is Tiger's intimidating presence. However, differences could still be due to how the course was set up or the weather or any number of factors (including random variation).

Brown divided her data into periods where Tiger's play is dominant (winning almost everything), typical (winning some), and struggling (not winning). The results are provocative. Brown finds that, during periods when Tiger is struggling, there is no significant difference in players' performance either with or without Tiger. During typical periods, the top players perform almost a stroke worse when Tiger is in the field. During dominant periods, the drops in performance average nearly 2 strokes per tournament. Brown's analysis supports her hypothesis that the presence of a "sure" winner can serve as a disincentive for other competitors.[7]

It would be interesting to repeat Brown's analysis in tournaments with a runaway leader who is not Tiger. Thinking about how to define a "runaway leader" presents some of the difficulties in knowing whether a study of this sort yields useful information. If someone is unchallenged and wins by several strokes, does that mean that the rest played worse than usual? I would say yes. After all, if any of the trailers played well, the runaway leader would not be likely to win by much. But did they play worse because they had little chance of winning, or did they have little chance of winning because (for some other reason) they played poorly? Here is another way of asking the same question. Nick Faldo won three Masters tournaments after dramatic collapses by very good golfers (Scott Hoch, Raymond Floyd, and Greg Norman). Does this mean that there was a "Faldo effect," or am I asking about a Faldo effect because he got lucky three times?

The Back Tee: Two Routes to the Same Hole

The win/loss and strokes rating systems described in this chapter turn out to be equivalent to a *least squares* system for predicting individual tournaments. Some details are given here.

To make the discussion less abstract, let's imagine a tour consisting of four golfers (A, B, C, and D) and three tournaments (T1, T2, and T3). Player A skipped tournament T3, and the tournament results were:

T1		T2		T3	
A:	E	A:	E	B:	E
B:	+1	C:	+4	C:	+9
C:	+9	B:	+5	D:	+12
D:	+11	D:	+8		

The strokes system defines an equation for the golfers' ratings. We denote the ratings by a, b, c, and d. In any tournament, the ratings predict player A to beat players B, C, and D by $a - b$ strokes, $a - c$ strokes, and $a - d$ strokes, respectively. In each of tournaments T1 and T2, then, player A should be ahead by $a - b + a - c + a - d = 3a - b - c - d$ strokes. For the two tournaments, player A should be ahead by twice this, or $6a - 2b - 2c - 2d$ strokes. In actuality, player A was ahead by $1 + 9 + 11 + 4 + 5 + 8 = 38$ strokes. Player A's equation is therefore $6a - 2b - 2c - 2d = 38$, and we repeat the process for each player. The other equations are $8b - 2a - 3c - 3d = 35$, $8c - 2a - 3b - 3d = -20$, and $8d - 2a - 3b - 3c = -53$.

The equations can be represented in a matrix:

$$\begin{bmatrix} 6 & -2 & -2 & -2 & 38 \\ -2 & 8 & -3 & -3 & 35 \\ -2 & -3 & 8 & -3 & -20 \\ -2 & -3 & -3 & 8 & -53 \end{bmatrix},$$

which can be reduced to

$$\begin{bmatrix} 1 & 0 & 0 & -1 & 10 \\ 0 & 1 & 0 & -1 & 8 \\ 0 & 0 & 1 & -1 & 3 \\ 0 & 0 & 0 & 0 & 0 \end{bmatrix}.$$

The reduced matrix translates to the equations $a-d=10$, $b-d=8$, $c-d=3$, and $0=0$. Interestingly, there is an infinite number of solutions to our system of equations. To generate a particular solution, choose a value for d and solve for the other three ratings. A simple choice is $d=0$, which leads to $c=3$, $b=8$, and $a=10$. This says that player A is 2 strokes better than player B, who is 5 strokes better than player C, who is 3 strokes better than player D. Fortunately, all solutions give the same relative rankings and the same stroke differences between players. In this analysis, the equations were the result of insisting that each player's *predicted* net strokes exactly match the player's *actual* net strokes for the season. Individual tournament results can vary from the predictions, but the season totals must match.

A different way of approaching the rating system is to look at each individual tournament. For example, in T1 we predict player A to beat player B by $a-b$ strokes. The actual result is player A winning by 1 stroke. The difference between the predicted result and the actual result is $(a-b)-1$. We interpret this as the error in the prediction. The error for A versus C in T1 is $(a-c)-9$, the error for A versus D in T1 is $(a-d)-11$, and so on. Our goal is to choose a, b, c, and d to make the errors as small as possible. Of the various ways in which we might define "as small as possible," we choose to minimize the sum of the squares of the errors. That is, we want to minimize $\left[(a-b)-1\right]^2 + \left[(a-c)-9\right]^2 + \left[(a-d)-11\right]^2 + \ldots$ for all of the two-player match-ups for the season.

Calculus gives us a straightforward way of finding the smallest sum. You compute partial derivatives of the function to be minimized with respect to each of the four variables and set all derivatives equal to 0. This produces four equations that turn out to be equivalent to the equations for the strokes system. The strokes rating system (and, by analogy, the win/loss rating system) can therefore be thought of in two ways. In the first, the ratings are produced by requiring the predicted results and actual results for the entire season to match, while in the second, the ratings are produced by making the results in each of the individual tournaments match the predictions as closely as possible. The rating system is therefore sound on the large scale (the entire season) and the small scale (individual tournaments).

Appendix A. Supplementary Tables

Chapter Seven

Table A7.1 Percentages of putts made from 11–15 feet

| | Distance (ft) | | | | |
	11	12	13	14	15
Birdie	.357	.313	.274	.254	.231
Par	.388	.338	.309	.302	.278
Bogey	.402	.365	.280	.280	.351

Table A7.2 Percentages of putts made from 11–15 feet

| | Distance (ft) | | | | |
	11	12	13	14	15
1st	.361	.316	.282	.262	.241
2nd	.516	.443	.355	.418	.464

Table A7.3 Percentages of first putts made from 11–15 feet

| | Distance (ft) | | | | |
	11	12	13	14	15
1st Birdie	.357	.313	.275	.254	.232
1st Par	.368	.325	.302	.290	.265

Chapter Eight

Table A8.1a Top ten average approach distances for shots made from 4–50 yards, 2006

		Fairway		Rough	
Rank	Player	Distance (ft)	Player	Distance (ft)
1.	E. Els	2.90	T. Herron	3.43
2.	S. Stricker	2.54	O. Uresti	3.32
3.	C. Barlow	2.51	S. Stricker	3.02
4.	P. Mickelson	2.25	S. Ames	2.88
5.	M. Weir	2.20	B. Andrade	2.84
6.	D. Love	2.08	C. Riley	2.45
7.	S. Verplank	2.08	P. Harrington	2.33
8.	R. Pampling	1.98	J. Rose	2.32
9.	B. Quigley	1.93	P. Perez	2.31
10.	V. Singh	1.91	K. Na	2.24

Table A8.1b Top ten average approach distances for shots made from 4–50 yards, 2007

		Fairway		Rough	
Rank	Player	Distance (ft)	Player	Distance (ft)
1.	P. Mickelson	3.04	K. J. Choi	3.42
2.	S. Stricker	3.02	R. Sabbatini	3.05
3.	S. Garcia	2.99	S. Stricker	3.02
4.	P. Goydos	2.39	S. Appleby	2.98
5.	E. Els	2.35	K. Triplett	2.96
6.	P. Harrington	2.31	K. Sutherland	2.94
7.	C. Howell	2.29	J. Furyk	2.85
8.	R. Imada	2.22	V. Singh	2.82
9.	J. Wagner	2.12	I. Poulter	2.81
10.	I. Poulter	2.07	L. Janzen	2.64

Table A8.1c Top ten average approach distances for shots made from 4–50 yards, 2008

		Fairway		Rough	
Rank	Player	Distance (ft)	Player		Distance (ft)
1.	K. J. Choi	2.86	T. Woods		3.87
2.	P. Harrington	2.63	J. Coceres		3.62
3.	T. Woods	2.33	C. Riley		3.23
4.	S. Stricker	2.25	J. Kaye		3.17
5.	T. Wilkinson	2.09	T. Immelman		2.97
6.	S. Maruyama	2.02	M. Brooks		2.31
7.	T. Immelman	1.92	L. Mattiace		2.30
8.	F. Couples	1.84	J. Leonard		2.20
9.	M. Weir	1.84	B. Gay		2.19
10.	S. Garcia	1.78	S. Stricker		2.17

Table A8.2 Top ten average approach distances for shots made from greenside bunkers, various distances from tide, 2004–2008

	Approach shot made from 0–30 feet		Approach shot made from 90–120 feet		Approach shot made from 120–150 feet	
Rank	Player	Distance (ft)	Player	Distance (ft)	Player	Distance (ft)
1.	J. Kelly	5.21	M. O'Meara	9.03	J. Leonard	8.82
2.	D. Pride	5.84	K. Triplett	9.97	O. Browne	9.38
3.	J. Furyk	5.85	R. Imada	10.92	J. Huston	9.86
4.	C. Pavin	6.21	S. Appleby	11.91	J. Driscoll	11.17
5.	L. Donald	6.21	M. Weir	12.08	O. Uresti	11.67
6.	V. Singh	6.26	S. Flesch	12.28	S. O'Hair	11.96
7.	P. Stankowski	6.29	V. Singh	12.37	J. Rose	12.18
8.	R. Pampling	6.53	D. Hart	12.50	R. Damron	12.97
9.	O. Uresti	6.59	M. Allen	12.62	A. Oberholser	13.30
10.	R. Damron	6.62	D. Forsman	12.65	S. Garcia	13.50
	Average	9.52	Average	17.44	Average	23.49

Chapter Nine

Table A9.1 Top ten average approach distances for shots made from the rough, various distances, 2008

| | Approach shot made from 75–100 yards | | Approach shot made from 100–125 yards | | Approach shot made from 125–150 yards | |
| | | Distance | | Distance | | Distance |
Rank	Player	(ft)	Player	(ft)	Player	(ft)
1.	B. Gay	13.8	F. Funk	15.9	J. Kaye	28.1
2.	R. Imada	14.8	L. Mattiace	16.1	A. Buckle	31.9
3.	T. Woods	15.6	A. Canizares	18.8	J. L. Lewis	32.1
4.	C. Collins	16.7	C. Stroud	20.6	A. Cejka	32.9
5.	B. Jobe	16.9	H. Slocum	22.1	S. Allan	33.5
6.	G. Kraft	17.4	K. Ferrie	22.1	P. Casey	34.6
7.	N. O'Hern	17.5	N. Begay	22.8	B. Pappas	34.9
8.	S. Bertsch	18.1	S. Micheel	23.6	B. Curtis	35.7
9.	D. Pride	18.7	J. Kelly	24.1	T. Clark	36.2
10.	P. Harrington	18.7	D. Pride	24.4	V. Singh	36.5
	Average	33.8	Average	43.2	Average	56.3

Chapter Ten

Table A10.1 Top ten ratings in driving accuracy, 2007 and 2008

| | | 2007 | | 2008 |
| | | Fairways | | Fairways |
Rank	Player	hit (%)	Player	hit (%)
1.	J. Van Zyl	79.5	O. Browne	80.4
2.	P. Goydos	78.7	F. Funk	77.0
3.	T. Armour	78.3	L. Mize	76.7
4.	M. Boyd	77.9	O. Uresti	74.8
5.	Z. Johnson	76.7	M. Brooks	74.8
6.	F. Funk	76.4	T. Byrum	74.2
7.	C. Bowden	76.1	H. Slocum	73.9
8.	M. Brooks	75.6	B. Bryant	73.9
9.	C. Beckman	75.3	Z. Johnson	73.7
10.	D. LaBelle	75.3	S. Verplank	73.6
	Average	63.0	Average	63.2

Chapter Eleven

Table A11.1 Putting efficiency ratings: top ten putts per round better than Tour average, from the same distances, 2004–2006

	2004		2005		2006	
Rank	Player	Rating	Player	Rating	Player	Rating
1.	L. Mize	1.01	B. Crane	1.02	B. Crane	0.94
2.	T. Woods	0.93	S. Stricker	0.80	S. Cink	0.83
3.	A. Scott	0.88	B. Faxon	0.75	R. Pampling	0.76
4.	B. Faxon	0.86	B. Tway	0.74	J. Furyk	0.75
5.	G. Chalmers	0.82	D. Clarke	0.72	R. S. Johnson	0.75
6.	S. Cink	0.80	L. Mattiace	0.71	B. Gay	0.74
7.	S. Stricker	0.79	A. Atwal	0.71	B. Quigley	0.71
8.	B. Geiberger	0.74	T. Woods	0.70	D. Wilson	0.68
9.	L. Roberts	0.73	J.-M. Olazabal	0.68	D. Chopra	0.67
10.	A. Baddeley	0.71	S. Cink	0.67	J. Parnevik	0.66

Table A11.2 Strokes per shot better than Tour average, from bunker

	2008		2009		2004–2009	
Rank	Player	Rating	Player	Rating	Player	Rating
1.	M. Weir	0.147	R. Allenby	0.124	M. Weir	0.101
2.	O. Uresti	0.134	C. Riley	0.104	M. Wilson	0.086
3.	N. Lancaster	0.129	J. Oh	0.102	N. Price	0.085
4.	A. Scott	0.128	O. Wilson	0.091	M. Dawson	0.080
5.	B. Rumford	0.125	I. Poulter	0.091	C. Riley	0.079
6.	P. Mickelson	0.118	A. Scott	0.087	T. Clark	0.078
7.	J. Gove	0.114	M. Weir	0.082	L. Donald	0.074
8.	C. Howell	0.107	R. Imada	0.082	J. Sluman	0.069
9.	K. Na	0.098	N. O'Hern	0.081	K. J. Choi	0.067
10.	A. Oberholser	0.096	K. J. Choi	0.076	R. Sabbatini	0.065

Table A11.3 Strokes per drive better than Tour average, par-5 holes

Rank	2008			2009			2004–2009		
	Player	Rating		Player	Rating		Player	Rating	
1.	B. Watson	0.105		T. Ridings	0.109		B. Watson	0.077	
2.	R. Garrigus	0.100		K. Stanley	0.107		J. Holmes	0.070	
3.	N. Watney	0.096		T. Woods	0.092		R. Garrigus	0.063	
4.	J. Holmes	0.096		R. Fisher	0.089		A. Kim	0.060	
5.	S. Allan	0.088		G. Woodland	0.088		A. Cabrera	0.059	
6.	J. Gore	0.071		R. Garrigus	0.088		N. Watney	0.055	
7.	C. Warren	0.063		M. Laird	0.087		S. Gutschewski	0.053	
8.	T. Ridings	0.063		R. Fowler	0.076		C. Warren	0.051	
9.	T. Gainey	0.062		D. Johnson	0.076		B. Wetterich	0.050	
10.	D. Johnson	0.060		T. Gainey	0.071		B. Weekley	0.049	

Table A11.4 Second ten overall strokes better than Tour average, 2007–2009

Rank	2007			2008			2009		
	Player	Rating		Player	Rating		Player	Rating	
11.	V. Singh	1.08		R. Palmer	0.94		D. Toms	1.05	
12.	A. Baddeley	1.08		B. Crane	0.91		J. Leonard	1.00	
13.	S. O'Hair	1.05		K. Sutherland	0.90		R. Fowler	0.99	
14.	A. Oberholser	1.02		T. Clark	0.89		S. O'Hair	0.95	
15.	J. Furyk	1.02		B. Gay	0.84		K. Perry	0.94	
16.	H. Slocum	1.01		P. Perez	0.83		P. Mickelson	0.92	
17.	J. Rose	1.01		S. Verplank	0.82		M. Weir	0.91	
18.	B. Snedeker	0.92		J. Furyk	0.78		C. Hoffman	0.91	
19.	L. Glover	0.92		C. Villegas	0.78		B. Gay	0.90	
20.	P. Harrington	0.91		S. Garcia	0.78		B. Snedeker	0.87	

Appendix B. Player Profiles

My purpose in creating player profiles for 15 golfers is to see if the numbers have a story to tell. My conclusions about strengths and weaknesses are based solely on the numerical ratings, so it would be interesting to compare my data-driven conclusions to those of the experts, the players who play the Tour, and the coaches, reporters, and others who watch and analyze every aspect of golf.

There were several criteria for selection. All players were in my data set for each year from 2004 to 2009 (except for John Daly in 2009). I took the dominant players of the era—Tiger Woods, Phil Mickelson, Steve Stricker, Vijay Singh, and Ernie Els—and asked, What makes them the best? I also included players who had been in and out of the top ten and wondered, Can we draw any conclusions about which skills cause players to jump into or drop out of the top echelon? Finally, I wanted a few solid players who are sometimes in contention and have won tournaments but are never mentioned as likely winners of a major. Can we identify characteristics that separate them from the top players?

For each player and each year, I list their rating and rank (using the Total Strokes system described in chapter 11) and then each of the components of the rating system. For example, values in the column labeled "L-F" correspond to the players' ratings hitting from the fairway from 50 to 200 yards out. (Top tens for the raw ratings are given in chapter 11, and complete lists are at www.roanoke.edu/mcsp/minton/ShotLink.html.) The entry here reflects the actual contribution of that category toward the rating. That is, I assumed that the average player hits 10 such shots, so the actual rating at 50 to 200 yards is multiplied by 10 in the Total Strokes formula. This multiplied rating (rounded to two digits) is shown in the tables. Thus, we can see that in 2008 Tiger gained 0.71 strokes per round on the average pro with his excellent approach shots from 50 to 200 yards from the fairway, plus another 0.12 strokes from the rough. You can identify strengths and weaknesses for a given year by looking left to right, and then look top to bottom to see if these strengths and weaknesses are repeated over time or were one-year anomalies.

Remember that the ShotLink data are limited to PGA Tour events and exclude the majors and overseas tournaments. For this reason, golfers such as Ernie Els and Retief Goosen who frequently play outside the United States may be misrepresented here. Much of their golfing year does not register in my data sets. This also explains why Padraig Harrington, on whom I had very little data for 2004 and 2005, is not in this appendix.

LEGEND

Rating	= total rating
Putt	= putting rating for round
T-3	= rating for 4 par-3 tee shots
T-4	= rating for 11 par-4 tee shots
T-5	= rating for 3 par-5 tee shots
L-F	= rating for 10 approach shots of 50–200 yards from the fairway
L-R	= rating for 4 approach shots of 50–200 yards from the rough
S-F	= rating for 3 chips of 4–50 yards from the fairway
S-R	= rating for 3 chips of 4–50 yards from the rough
Par5	= rating for 2 shots of 200–250 yards from the fairway
Sand	= rating for 1 shot from the sand

Tiger Woods

Year	2009	2008	2007	2006	2005	2004
Rating	2.72	2.58	2.72	2.21	2.21	2.19
Rank	1	1	1	1	1	1

Year	Putt	T-3	T-4	T-5	L-F	L-R	S-F	S-R	Par5	Sand
2010	0.11	0.12	−.11	0.01	0.39	−.04	0.27	−.06	−.01	−.06
2009	0.99	0.21	0.16	0.28	0.42	−.03	0.38	0.12	0.13	0.06
2008	0.84	0.19	0.03	−.05	0.71	0.12	0.32	0.32	0.13	−.04
2007	0.75	0.26	0.16	0.11	0.66	0.41	0.01	0.21	0.11	0.04
2006	0.56	0.22	0.05	0.06	0.63	0.24	0.12	0.01	0.29	0.03
2005	0.70	0.12	0.33	0.20	0.49	0.16	0.03	−.03	0.16	0.06
2004	0.93	−.03	−.11	−.04	0.54	0.14	0.34	0.20	0.19	0.03

From 2004 to 2009, Tiger was the best at putting and approach shots from the fairway. He was very strong on par-3 tee shots, 200- to 250-yard approach shots, all shots from the rough, and short shots from the fairway. His driving and bunker play was above average. The most impressive aspect of his stats is the consistently high number of large values in the table. He was just the best at almost everything. Until 2010, that is. I include his numbers here to document his unusual year. His putting dropped from consistently outstanding to mediocre, and his driving went from extremely good in 2009 to awful. Between putting and tee shots on par 4s and 5s, he rated more than 1.4 strokes per round worse in 2010 than in 2009. After having only four individual skills ratings below zero from 2005 to 2009, Tiger had five negative ratings in 2010. That said, his overall rating was still 0.62 strokes better than average, ranking him 41st. Only in the context of Tiger Woods's career would we call this a disaster.

Phil Mickelson

Year	2009	2008	2007	2006	2005	2004
Rating	0.92	1.17	1.45	1.18	1.30	1.02
Rank	16	7	3	5	5	6

Year	Putt	T-3	T-4	T-5	L-F	L-R	S-F	S-R	Par5	Sand
2009	−.08	0.09	0.17	0.19	−.11	0.15	0.20	0.21	0.09	0.01
2008	0.23	0.12	0.06	0.13	0.35	0.01	0.08	0.04	0.03	0.12
2007	0.06	0.14	0.08	0.09	0.13	0.12	0.41	0.26	0.07	0.09
2006	0.03	0.13	0.06	0.02	0.31	0.13	0.29	0.00	0.22	−.03
2005	0.13	0.08	0.24	0.06	0.22	0.35	0.18	0.13	−.10	0.02
2004	−.08	0.06	0.28	−.14	0.33	−.05	0.34	0.02	0.20	0.06

Phil is the only player other than Tiger to make the top ten overall each year from 2004 to 2008, but he fell to 16th in 2009. The most dramatic difference between Phil and Tiger's numbers is in the putting column. Phil lags behind Tiger by over a half-stroke per round in putting efficiency. Like Tiger, Phil has very few negative entries. There are no noticeable deficiencies in his game, other than his strengths not being strong enough. His most impressive numbers are approach shots from the fairway, long and short. His wedge game, which is highly acclaimed, ranks near the top (categories S-F, S-R, and Sand).

Steve Stricker

Year	2009	2008	2007	2006	2005	2004
Rating	1.52	1.02	1.69	1.74	1.18	0.01
Rank	2	8	2	2	7	98

Year	Putt	T-3	T-4	T-5	L-F	L-R	S-F	S-R	Par5	Sand
2009	0.41	0.09	0.09	0.01	0.64	0.01	0.26	−.00	0.00	0.06
2008	0.46	0.03	−.21	−.06	0.36	−.04	0.28	0.22	−.03	0.02
2007	0.52	0.09	−.20	−.07	0.68	0.04	0.35	0.29	−.07	0.06
2006	0.58	0.13	−.02	−.03	0.28	0.15	0.36	0.26	−.03	0.07
2005	0.80	−.15	−.40	−.19	0.47	0.16	0.23	0.19	0.03	0.04
2004	0.79	−.13	−.65	−.34	0.12	0.02	0.10	0.17	−.07	0.01

Since bottoming out in 2004, Steve Stricker has been in the top ten every year. His top ten in 2005 is actually an interesting comment on the rating system. He had lost his playing card, so his 2005 record was compiled from the events for which he could qualify or receive a sponsor's exemption. He finished 162nd on the money list, having to go back to Q-School to earn back his playing card. In spite of his poor money showing in 2005, the ratings indicate that he was playing as well as anybody (except You Know Who). He is excellent at putting, on long approach shots from the fairway, and on short shots of any type. These strengths enable him to stay near the top in spite of relatively weak driving statistics. In 2004, his driving was regularly costing him a full stroke per round. When his driving is only a little negative or positive, he is tough. He would be a great Captain's Choice teammate, if you paired him with a great driver like J. B. Homes.

Vijay Singh

Year	2009	2008	2007	2006	2005	2004
Rating	−0.21	1.14	1.08	0.88	1.52	1.60
Rank	150	3	9	11	2	2

Year	Putt	T-3	T-4	T-5	L-F	L-R	S-F	S-R	Par5	Sand
2009	−.66	0.11	0.27	0.10	−.09	0.06	0.09	−.01	−.10	0.03
2008	−.41	0.15	0.42	0.16	0.19	0.11	0.20	0.24	0.03	0.05
2007	−.10	0.11	0.17	0.11	0.08	0.14	0.16	0.22	0.12	0.06
2006	0.04	0.10	−.05	0.00	0.19	0.16	0.26	0.13	−.04	0.09
2005	0.26	0.08	0.37	0.19	0.12	0.24	0.04	0.18	−.01	0.04
2004	0.01	0.10	0.45	0.05	0.49	0.18	0.08	0.07	0.09	0.06

In the early 2000s, Vijay challenged Tiger for the number-one slot. His drop-off from being second overall in 2004 and 2005 to borderline top ten in 2006 and 2007 was (numerically, at least) due to driving problems. He was a half stroke or more above average in 2004, 2005, and 2008 for par-4 and par-5 tee shots and slightly below average in 2006. Vijay's putting was average until 2008, when he started losing half a stroke per round to the field. His number-three ranking in 2008, while giving away four-tenths of a stroke on the greens, is remarkable. Vijay is consistently impressive with an iron or wedge in his hands but had an across-the-board drop in performance in 2009.

Ernie Els

Year	2009	2008	2007	2006	2005	2004
Rating	0.58	0.23	1.39	1.13	1.31	1.01
Rank	42	90	5	6	5	11

Year	Putt	T-3	T-4	T-5	L-F	L-R	S-F	S-R	Par5	Sand
2009	−.35	0.13	0.11	0.02	0.29	0.04	0.34	−.07	0.03	0.03
2008	−.40	0.17	−.08	−.08	0.43	−.06	0.05	0.05	0.10	0.06
2007	0.07	0.22	0.09	0.10	0.22	0.11	0.36	0.00	0.20	0.02
2006	0.01	0.12	0.00	0.03	0.35	−.06	0.37	0.09	0.15	0.06
2005	0.27	0.17	0.23	0.10	0.45	0.13	−.18	0.04	0.14	−.04
2004	0.18	0.30	−.07	−.01	0.10	−.03	0.27	0.06	0.15	0.07

Ernie is another player who at one time was considered to be Tiger's equal. Iron shots from the tee and the fairway show up as strong points each year. His putting slid from above average to average to below average in 2008 and 2009. Most of his ratings were worse in 2008, but I don't know whether that was due to injuries, lack of confidence from not winning any tournaments in 2007 (even though his rating was high and he did well in the majors), or some other factor. His 2007 stats are all positive—one of the few examples in my records of someone rating above average in every category.

Sergio Garcia

Year	2009	2008	2007	2006	2005	2004
Rating	0.73	0.78	1.30	0.35	0.67	0.56
Rank	36	13	6	65	27	45

Year	Putt	T-3	T-4	T-5	L-F	L-R	S-F	S-R	Par5	Sand
2009	0.05	0.01	0.13	0.12	0.11	0.36	0.017	−.22	0.04	−.03
2008	−.05	0.06	0.13	−.01	0.18	0.10	0.20	0.04	0.08	0.03
2007	0.32	0.11	−.14	−.05	0.23	0.07	0.39	0.11	0.18	0.07
2006	−.29	0.18	0.02	0.03	0.15	0.17	0.03	−.13	0.20	−.01
2005	−.32	0.15	0.35	0.08	0.09	0.05	0.12	−.04	0.12	0.05
2004	−.53	0.08	0.16	−.13	0.55	0.26	0.05	0.04	0.04	0.05

The six years covered by my data record one cycle in Sergio's oscillation between elite player and good player. His top ten in 2007 followed several lackluster years, during which many doubted whether he would ever fulfill his enormous potential. His rating as a putter is very unusual. From putting ratings that are well below average (his rating in 2004 was 1.5 strokes per round worse than Tiger's; that's a lot of strokes to give away), he suddenly had a very good year on the greens in 2007. His stroke-per-round improvement from 2006 to 2007 came on the greens and fairway approach shots from 4–50 yards. Hit it close and make the putt!

Jim Furyk

Year	2009	2008	2007	2006	2005	2004
Rating	1.27	0.78	1.02	1.52	1.42	1.00
Rank	4	14	11	3	3	12

Year	Putt	T-3	T-4	T-5	L-F	L-R	S-F	S-R	Par5	Sand
2009	0.81	0.09	0.08	−.06	0.33	−.29	0.32	−.03	0.00	0.03
2008	0.36	0.03	0.09	−.06	0.22	−.05	0.14	0.04	0.03	−.02
2007	−.05	0.07	0.17	−.11	0.34	−.11	0.25	0.30	0.05	0.10
2006	0.75	0.06	−.00	−.01	0.36	−.05	0.06	0.23	0.12	−.00
2005	0.47	0.14	−.01	−.08	0.58	−.20	0.27	0.10	0.06	0.09
2004	0.58	0.03	0.28	−.19	0.27	−.14	0.20	−.08	0.07	−.03

The guy with the funny swing can play! Jim Furyk is known as someone who gets all he can out of his game. His statistics do not vary much from year to year, an indication that he works hard and concentrates well. His fall out of the top ten is attributable to putting woes. He suddenly dropped from being one of the top putters to average performance in 2007, then up to good in 2008, and back to excellent in 2009. He has two consistently negative categories—tee shots on par 5s and long approach shots from the rough. Strength is not his strength.

K. J. Choi

Year	2009	2008	2007	2006	2005	2004
Rating	0.21	1.02	1.08	0.15	0.37	−0.14
Rank	89	5	8	97	68	147

Year	Putt	T-3	T-4	T-5	L-F	L-R	S-F	S-R	Par5	Sand
2009	−.03	0.06	0.02	−.06	−.13	0.06	0.16	0.04	0.01	0.08
2008	0.58	−.04	−.05	0.05	0.14	−.11	0.33	0.03	0.06	0.04
2007	0.60	0.01	−.01	−.02	−.02	−.06	0.13	0.26	0.07	0.13
2006	0.25	0.02	−.03	−.01	0.08	−.11	0.03	−.14	0.03	0.04
2005	0.21	−.00	0.10	−.01	0.00	−.02	−.05	0.05	0.05	0.03
2004	0.21	−.00	−.11	−.18	0.23	−.05	−.14	−.08	−.11	0.09

Choi emerged in 2007 as a top ten player. As with most players in this group who show a large change in ranking from one year to the next, Choi's most notable area of change was in putting. Good putting ratings from 2004 to 2006 suddenly became top notch, and K. J. suddenly started winning important tournaments. His other short game stats also improved noticeably for 2007. However, his putting suddenly became average in 2009, accounting for much of the drop in his rating. Aside from putting and shots from inside of 50 yards, his ratings are most noteworthy for not being noteworthy: he rates as basically average in all other areas.

John Daly

Year	2008	2007	2006	2005	2004
Rating	−0.85	−0.24	−0.48	0.02	0.98
Rank	212	153	167	106	12

Year	Putt	T-3	T-4	T-5	L-F	L-R	S-F	S-R	Par5	Sand
2008	−.88	−.21	0.18	0.15	0.16	−.12	0.04	−.07	−.09	0.00
2007	−.37	0.05	0.15	0.04	−.42	0.14	0.07	−.03	0.16	−.03
2006	−.54	0.00	0.05	0.05	−.26	0.09	0.15	−.04	0.02	−.01
2005	−.23	−.02	0.27	0.17	−.15	0.08	0.03	−.16	0.02	0.02
2004	0.16	−.04	0.23	−.01	0.25	0.08	0.14	−.03	0.14	0.05

Everybody's favorite barefoot, beer-bellied star golfer has fallen off the charts, but he was still playing well in 2004. I thought it would be interesting to check out his strengths (par-4 tee shots, long and short shots from the fairway) and weaknesses (none) when he was at the top of his game. Also, I wanted to see if he had deteriorated in all ways or only in isolated parts of his game. Putting jumps out: from above average in 2004 to bad in 2008, he lost a shot per round on the greens. He also lost some of his edge off the tees, and his iron game collapsed.

Justin Leonard

Year	2009	2008	2007	2006	2005	2004
Rating	1.00	1.18	0.68	−0.15	0.44	0.49
Rank	12	6	34	147	46	51

Year	Putt	T-3	T-4	T-5	L-F	L-R	S-F	S-R	Par5	Sand
2009	0.20	−.02	−.01	−.00	0.45	0.01	0.33	0.05	−.00	−.01
2008	0.55	−.05	0.12	−.07	0.35	0.03	0.09	0.17	−.04	0.03
2007	0.15	0.01	−.02	−.06	0.25	−.05	0.17	0.21	−.07	0.09
2006	−.11	−.16	−.03	−.01	0.16	−.17	0.07	0.14	−.09	0.04
2005	0.10	0.04	−.09	−.06	0.26	0.01	0.18	−.01	−.01	0.01
2004	−.11	0.05	−.01	−.17	0.43	0.01	0.06	0.13	0.06	0.05

Justin Leonard gives us a case study of someone who dropped down in the rankings (to 147th in 2006) and then came back. Again, one of my questions is whether the causes of decline (and subsequent recovery) can be found in all parts of his game or only in specific areas. Putting is not the primary culprit, for once. A big improvement in putting in 2008 elevated him to number 6, but his below-average putting score in 2006 matched his rating in 2004. If you look at the categories that measure iron play (T-3, L-F, and L-R), you'll see that he dropped from a total of 0.31 in 2005 to −0.17 in 2006, accounting for most of the half-stroke per round he dropped overall. In 2007, he bounced back to 0.21, and then in 2008 he further improved to 0.33. Putting is always important, but Justin gives us an example of iron play being critical. In 2009, it was also critical for him to hit the fairway, as his iron play from the fairway was exceptional while from the rough it was average.

Retief Goosen

Year	2009	2008	2007	2006	2005	2004
Rating	0.56	−0.08	−0.61	0.61	1.07	1.29
Rank	44	128	207	30	14	4

Year	Putt	T-3	T-4	T-5	L-F	L-R	S-F	S-R	Par5	Sand
2009	0.45	0.04	0.02	0.05	−.08	0.03	−.02	−.04	0.08	0.03
2008	−.11	0.04	−.15	−.15	0.13	−.18	0.08	0.22	0.03	0.01
2007	−.10	0.03	−.30	−.03	−.04	−.27	0.01	0.13	−.01	−.02
2006	0.00	0.04	0.01	0.02	0.42	−.13	0.19	0.00	−0.2	0.07
2005	0.09	0.15	0.29	0.06	0.16	0.01	0.00	0.16	0.16	0.01
2004	0.49	0.12	0.04	−.09	0.19	0.20	0.08	0.19	0.06	0.01

My main image of Retief Goosen is of a golfer winning the U.S. Open by making every putt in sight. It was, therefore, surprising to see him rate below average as a putter in all years except for 2004, when he was winning his second Open, and again in 2009, when his resurgence was fueled almost exclusively by better putting. Along with putting, the numerical explanation for Goosen's decline can be found in driving and approach shots from 50–200 yards. (Of course, add in putting and that's most of the game.) His driving dropped from average (2004) and well above average (2005) to well below average in 2007 and 2008, before returning to average in 2009.

Scott Verplank

Year	2009	2008	2007	2006	2005	2004
Rating	0.75	0.82	0.78	0.83	0.40	0.93
Rank	31	18	32	15	73	19

Year	Putt	T-3	T-4	T-5	L-F	L-R	S-F	S-R	Par5	Sand
2009	0.16	0.08	0.00	0.01	0.32	0.09	0.14	−.11	0.03	0.03
2008	−.27	0.06	0.13	−.01	0.43	0.24	0.08	0.05	0.03	0.08
2007	−.20	0.17	0.00	−.06	0.48	0.07	0.14	0.12	−.02	0.07
2006	0.19	0.15	−.05	−.06	0.27	−.01	0.23	0.08	0.02	−.00
2005	0.06	0.08	0.02	−.06	0.50	−.17	0.00	−.03	−.00	0.00
2004	0.37	0.05	0.19	−.19	0.58	−.10	0.02	0.01	−.02	0.03

Scott Verplank is as steady as you would expect watching his swing. Year in and year out, he has been around the second group of 30 players in the rankings. His dip in 2005 was due to putting. Interestingly, his putting has never returned to its 2004 level, but he has compensated with improvements in other areas, especially performance from the rough. He is not strong with the driver but is consistently excellent from 50–200 yards.

Mike Weir

Year	2009	2008	2007	2006	2005	2004
Rating	0.91	0.67	−0.06	0.50	0.01	0.42
Rank	17	23	160	43	94	55

Year	Putt	T-3	T-4	T-5	L-F	L-R	S-F	S-R	Par5	Sand
2009	0.56	0.04	−.18	−.14	0.31	−.06	0.22	0.13	−.06	0.08
2008	0.20	0.03	−.03	−.08	0.36	−.15	0.16	0.08	−.05	0.15
2007	−.41	0.04	0.15	0.01	0.07	−.12	−.08	0.20	−.05	0.12
2006	−.01	0.05	−.02	−.04	0.39	−.11	0.14	0.06	−.06	0.10
2005	−.20	−.04	0.01	−.06	−.07	−.18	0.22	0.25	−.02	0.10
2004	0.17	0.17	−.06	−.15	0.21	−.08	0.08	−.06	0.09	0.05

Mike Weir's numbers show a surprising amount of variation. His putting rating changed by at least one-third of a stroke per round most years. His overall rating tracks his putting rating nicely. His rating from the fairway at 50–200 yards was on an every-other-year cycle of good and average until a second straight good year in 2009. His short game ratings also fluctuate a fair amount, with the exception of a sterling bunker rating year after year. He is not strong with the driver or out of the rough from 50–200 yards. Like Kenny Perry, he had a heroic international event as "host": the 2007 Presidents Cup in Montreal.

John Rollins

Year	2009	2008	2007	2006	2005	2004
Rating	−0.14	−0.05	0.38	0.12	−0.22	0.58
Rank	142	139	78	110	152	27

Year	Putt	T-3	T-4	T-5	L-F	L-R	S-F	S-R	Par5	Sand
2009	0.06	−.14	0.31	0.10	−.22	−.13	0.03	−.17	0.01	−.00
2008	0.08	0.04	0.12	0.10	−.28	−.02	−.21	0.07	0.07	−.04
2007	0.12	−.06	0.16	0.05	−.08	0.00	0.02	0.06	0.05	0.05
2006	0.01	0.03	0.04	0.03	−.05	0.00	−.07	0.11	0.06	−.05
2005	−.31	−.04	0.19	0.13	−.24	0.03	−.00	−.01	0.06	−.04
2004	0.29	0.01	0.24	−.08	0.04	0.04	0.16	−.12	0.03	−.03

As Virginia Commonwealth University's most famous golfing alumnus, John Rollins has had a solid career. He reached a peak ranking of 27th in 2004, when he had his best putting performance, but he has a lot of ratings that are near zero; he rates as an average golfer in most categories. Rollins is above average with the driver, but an above-average rating on long 200- to 250-yard fairway shots is balanced by below-average ratings from 50–200 yards. He is an example of a golfer who plays often and well, but rarely breaks out of the pack with a spectacular performance.

Kenny Perry

Year	2009	2008	2007	2006	2005	2004
Rating	0.94	0.29	0.58	−0.29	0.88	0.11
Rank	15	92	56	180	17	114

Year	Putt	T-3	T-4	T-5	L-F	L-R	S-F	S-R	Par5	Sand
2009	0.37	0.17	0.29	0.17	−.02	−.07	0.13	−.04	−.02	−.02
2008	0.10	0.14	0.14	0.12	0.07	−.03	−.19	−.07	−.00	0.01
2007	−.24	0.11	0.28	0.10	0.13	0.13	0.01	0.11	0.04	−.07
2006	−.34	0.18	0.06	0.04	0.07	0.00	−.14	−.09	−.02	−.06
2005	−.12	0.28	0.33	0.21	0.14	0.11	0.05	−.14	−.00	0.02
2004	−.13	0.05	0.32	−.01	0.15	−.14	−.15	−.06	0.05	0.03

Kenny Perry is someone I would like to study more. My impression as a casual fan is that he can get extremely hot and win tournaments, but the evidence is that he often plays indifferent golf. I would be interested to see if there is numerical evidence of this. Perry is good with the driver and irons, but not so good (overall) with the putter or wedges. His only statistically strong year for putting was 2009, when he nearly won the Masters. He is a rare example in the data set of someone who raised his ranking (to 17th in 2005) almost entirely with good driving. He rated 0.82 on tee shots in 2005, with an overall rating of 0.88. In 2009, he rated 0.63 on tee shots, with an overall rating of 0.94. My hypothesis is that, when he putts and chips well, he is a force. As of 2010 Perry holds the record for most money made on the PGA Tour without winning a major. His performance in the 2008 Ryder Cup was special. He skipped the major championships to concentrate on qualifying for the Ryder Cup, which was held in his home state of Kentucky. He scored 2.5 points and was an emotional leader of the United States, which won, 16.5–11.5, for the first time since 1999.

Notes

Preface

1. James and his ideas are featured in the controversial *Moneyball* by Michael Lewis. *The Numbers Game* by Alan Schwarz gives an enjoyable and evenhanded history of baseball's statistics and the role of Bill James in changing the game.

2. The 2006 Joint Mathematics Meetings in San Antonio. In an MAA-sponsored Session on Mathematics of Sports and Games, William Branson of St. Cloud State University gave a talk titled "Bill James as an Exemplar of Statistical Writing."

3. Coauthored with Bob Smith of Millersville University of Pennsylvania. As I write, various versions of the fourth edition wait impatiently on my desk.

4. See Cook (2010), p. 195.

5. Don Wade (2001) includes, in his *Talking on Tour* compilation, a story told by Trevino. Tenison gets more airtime in Michael Bohn's *Money Golf* and in articles like *Golf Digest*'s dialogue with Raymond Floyd.

6. See R. H. Coop (1998).

7. Gummer (2009) tells the story of the remarkable Homer Kelley. I do not have the dedication to tackle Kelley's book on my own, but golfers such as Steve Elkington and Bobby Clampett give testimony that the man knew what he was talking about.

8. See Ayres (2007), p. 10.

9. Steve Evans, Senior VP of Information Systems at PGA Tour, Stephanie Chvala, and Mike Vitti were especially helpful in working with me on the statistical questions.

10. Thanks to my friend Bob Schultz for this wonderful phrase.

Chapter One. The Shape of Golf

Epigraph: Tiger Woods Learning Center, www.twlc.org/spotlight .html.

1. See http://www.exxonmobil.com/Corporate/community_ed_math _academy.aspx.

2. These include Werner and Greig's *How Golf Clubs Really Work*, Wishon's *The Search for the Perfect Golf Club*, and Zumerchik's *Newton on the Tee*. Other books of interest are listed in the reference section.

3. Details can be found in *Tommy's Honor*, Kevin Cook's book about Old Tom Morris and Young Tom Morris. Michael Bohn's *Money Golf* gives an enjoyable summary of golf's history, with a particular emphasis on betting. A fictionalized account of how golf was played in the 1400s is given in Bob Cupp's *The Edict*, which provides a possible explanation of James II's 1457 ban on golf in Scotland beyond the stated reason that Scots needed more time for archery practice.

The only hook that Newton is known to have suffered from was Robert Hooke. Hooke was Newton's elder, in a position to become his mentor as a leader of the Royal Society of London. Instead, Hooke developed a habit of loudly proclaiming that Newton's results were either inspired by Hooke, stolen from Hooke, or incorrect. This contributed greatly to Newton's aversion to publishing his results, which had a surprising influence on the development of science. See Jason Bardi's *The Calculus Wars* for details.

4. The magnitude of the drag force is proportional to the density of the air that the ball is in. The density of air in Denver is about 14% lower than at sea level, accounting for standard changes in temperature and pressure. The La Paz Golf Club in Bolivia is identified by Duncan Lennard's *Extreme Golf* as being the world's highest course at 10,650 feet. This altitude would reduce air drag by about 26%. The world's lowest golf course is Furnace Creek Resort in Death Valley, California, at 214 feet below sea level. Air density is one reason that golf here is a drag. The hottest course is Alice Springs Golf Club in Northern Territory, Australia. At its normal 125°F, the air density is 15% less than air at 30°F.

5. See note 4 above. Among the factors that affect air density and, therefore, air drag are *temperature* and *humidity*. Air drag is lower in high

temperatures, leading to those nice long drives in the summer. Some people find the humidity result counterintuitive: an increase in humidity *reduces* air drag, giving you even more distance on a hot, steamy day.

6. An exception to this rule is when the spin axis and velocity vector are parallel. In this case, there is no Magnus force. The Magnus force is represented mathematically by a cross product (see Smith and Minton, *Calculus*), and the cross product of parallel vectors is the zero vector. The most common sports situation to which the exception applies is a spiral in football. It is good news, indeed, that spirals do not slice or hook, or the forward pass may have been so uncontrollable that it would never have been fully incorporated into the game.

7. Thanks to Acushnet Company for permission to use these figures, which can be found at the Titleist website. www.titleist.com/technology /details.asp?id=20.

8. The parabolic shape is also dependent on the gravitational force being constant. For the flight of a golf ball, this is a reasonable assumption.

9. The "gravity only" curve in figure 1.4 shows the solution of the differential equations $x''(t) = 0$ ft/s^2, $y''(t) = -32$ ft/s^2 with initial conditions derived from a launch speed of 234 ft/s at angle 15° and initial position at the origin. That is, $x(0) = y(0) = 0$ ft, $x'(0) = 234 \cos(15°)$ ft/s and $y'(0) = 234 \sin(15°)$ ft/s. The "gravity plus drag" curve in figure 1.4 shows a numerical approximation of the following differential equations with the same initial conditions:

$$x''(t) = -0.0012743 \sqrt{x'(t)^2 + y'(t)^2} * x'(t),$$

$$y''(t) = -32 - 0.0012743 \sqrt{x'(t)^2 + y'(t)^2} * y'(t).$$

These equations are consistent with the model published by Smits and Smith in *Science and Golf II* and with the drag data published online by Titleist (shown in figure 1.3). In these equations, the acceleration due to drag has the form $c |\mathbf{v}| \mathbf{v}$, where $\mathbf{v}$ is the velocity vector and $c = 0.0012743$ ft^{-1}. A better model would use a nonconstant c with a dependence on speed. Mathematica software generates the numerical approximations shown.

10. The "gravity only" curve in figure 1.5 is the same as in figure 1.4. For the "gravity, drag, and Magnus" curve in figure 1.5, the same launch speed, angle, and initial position are used. The initial spin is backspin with

a magnitude of 3,000 rpm (more precisely, 100π radians per second). The differential equations are

$$x''(t) = [-0.0012743 \sqrt{x'(t)^2 + y'(t)^2} - 0.0000669\omega(t)] * x'(t)$$
$$- 0.000936 \, \omega(t)^{0.4} \left(x'(t)^2 + y'(t)^2\right)^{0.3} y'(t),$$

$$y''(t) = -32 + [-0.0012743\sqrt{x'(t)^2 + y'(t)^2} - 0.0000669\omega(t)] * y'(t)$$
$$+ 0.000936 \, \omega(t)^{0.4} \left(x'(t)^2 + y'(t)^2\right)^{0.3} x'(t), \text{ and}$$

$$\omega'(t) = -\omega(t)/20,$$

which were developed by Roanoke College student Geoff Boyer based on the model of Smits and Smith. The drag term is the same as before except for including a correction for spin. The general form is now $c_1 |v| v + c_2 |\omega| v$, where ω is the spin rate in radians per second and $c_2 = 0.0000669$. The acceleration due to spin has the form $c_3 s \times v$, where s is the spin vector (with magnitude ω). In this example, $s = <\omega, 0, 0>$. The scalar c_3 has the form $c_4 \left(\frac{|v|}{|\omega|}\right)^{0.6}$, where $c_4 = 0.000936$ ft$^{-0.6}$, showing that the acceleration due to the Magnus force depends on the spin rate and speed. The spin rate ω is assumed to decay exponentially with a characteristic time of 20 seconds. The spin decay used here is from Werner and Greig, *How Golf Clubs Really Work*, p. 121. A slightly different result can be found in Zagarola, Lieberman, and Smits, "An Indoor Testing Range to Measure the Aerodynamic Performance of Golf Balls," pp. 348–354.

11. The comparison in figure 1.5 is not entirely fair. While the launch angle of $15°$ is near the optimal launch angle for a ball with a spin of 3,000 rpm, it is not at all close to the optimal launch angle in a vacuum, which equals $45°$. At this angle, the ball would travel 1,711 feet, almost 580 yards!

12. The three-dimensional model is

$$x''(t) = [-0.0012743\sqrt{x'(t)^2 + y'(t)^2 + z'(t)^2} - 0.0000669\omega(t)] * x'(t)$$
$$-0.000936\omega(t)^{0.4}\left(x'(t)^2 + y'(t)^2 + z'(t)^2\right)^{0.3}[\omega_2 * z'(t) - \omega_3 * y'(t)],$$

$$y''(t) = [-0.0012743\sqrt{x'(t)^2 + y'(t)^2 + z'(t)^2} - 0.0000669\omega(t)] * y'(t)$$
$$+0.000936\omega(t)^{0.4}\left(x'(t)^2 + y'(t)^2 + z'(t)^2\right)^{0.3}[\omega_3 * x'(t) - \omega_1 * z'(t)],$$

$$z''(t) = -32 + [-0.0012743\sqrt{x'(t)^2 + y'(t)^2 + z'(t)^2} - 0.0000669\omega(t)]$$
$$* z'(t) + 0.000936\omega(t)^{0.4}(x'(t)^2 + y'(t)^2 + z'(t)^2)^{0.3}[\omega_1 * y'(t)$$
$$- \omega_2 * x'(t)], \text{ and}$$

$$\omega'(t) = -\omega(t)/20,$$

where $\omega(t) < \omega_1, \omega_2, \omega_3 >$ is the spin vector, with magnitude $\omega(t)$ and direction $< \omega_1, \omega_2, \omega_3 >$.

The initial conditions are $x(0) = y(0) = z(0) = 0$, $x'(0) = v \cos(\theta)$ $\sin(\beta)$, $y'(0) = v \sin(\theta) \cos(\beta)$, $z'(0) = v \sin(\alpha)$, and $\omega(0) = \omega$. For a given club loft angle A and swing speed S, the launch angle of the ball is approximated by

$$\theta = \tan^{-1}\left(\frac{1.67 \tan A}{0.5 \tan^2 A + 1.83}\right)$$

above vertical. For a swing with plane angle A_p and clubface angle A_f, the launch angle of the ball compared to the target ($\beta = 0$) is

$$\beta = 0.2A_p + 0.8A_f,$$

while the launch speed of the ball is given by

$$v = S\frac{1.83 \cos A}{\cos(A - \theta) + \frac{46}{180} \cos A \cos \theta},$$

and the initial spin rate is

$$\omega = [7.143S] \sin A.$$

Defining $\delta = \tan^{-1}(\sin(A_f - A_p)/\tan A)$, the initial spin vector has $\omega_1 = \cos(\delta)\cos(\beta)$, $\omega_2 = \cdot \cos(\delta)\sin(\beta)$, and $\omega_3 = -\sin(\delta)$.

13. For most golfers, the closed clubface for a hook also turns the clubface down, creating a lower launch angle. Similarly, the open clubface often is tilted upward, creating a higher launch angle. Very few aspects of the golf shot exist in isolation!

14. Calculations use the model of note 12, with a swing speed of 132 ft/s, a launch angle of 34°, and a coefficient of restitution of 0.65 (replacing the 0.83 used previously).

15. A launch angle of 42° produces a flat-terrain carry of 412.6 feet. A rise of 30 feet decreases the carry to 388.7 feet, and a drop of 30 feet increases the carry to 435.0 feet. A launch angle of 26° degrees produces a flat-terrain carry of 532.4 feet. A rise of 30 feet decreases the carry to 498.5 feet, and a drop of 30 feet increases the carry to 562.3 feet.

16. These calculations model a stronger player with a swing speed of 144 ft/s and a launch angle of 40°. The coefficient of restitution was increased to 0.7 so that the flat-terrain distances would match.

17. The flat-terrain distance is 482.8 feet, or about 160 yards. A rise of 30 feet decreases the carry to 460.5 feet, a change of 7.4 yards, and a drop of 30 feet increases the carry to 503.7 feet, a change of 6.9 yards.

18. Werner and Greig, pp. 59–60.

19. An equilibrium can be stable or unstable. "Stable" means that, if the current value is close enough to the equilibrium, it will get closer and closer until it is basically at the equilibrium. How close is "close enough" depends on the specific problem. A handy visual aid for stability is a pencil. Resting on its side it is at a stable equilibrium. If you drop the pencil from a position almost on its side, it will slam down and come to rest on its side. There is another equilibrium where the pencil is balanced on its point. This is "unstable"—if you miss the balance point by any amount, the pencil will fall and eventually end up at its stable equilibrium, resting on its side.

20. To be specific, the equilibrium was computed for different spin rates by setting the equations for $x''(t)$ and $y''(t)$ equal to 0 and solving for the two unknowns $x'(t)$ and $y'(t)$. For a spin rate of 527.25 radians per second, the equilibrium is approximately $x'(t) = 86.3$ ft/s and $y'(t) = -79.6$ ft/s. This converts to a slope of $\frac{dy}{dx} \approx \frac{-79.6}{86.3} \approx -0.92$.

21. In algebraic terms, the equilibrium line for this graph is a horizontal asymptote. This is a real example of a horizontal asymptote that has meaning, as well as a case in which the graph crosses the asymptote. (Most graphs in algebra class do not cross asymptotes.)

Chapter Two. Golfer's Spread

Epigraph: Downs Macrury, *Golfers on Golf,* p. 54.

1. The *Wikipedia* entry "Moe Norman" states, "Sam Snead . . . once described Norman as the greatest striker of the ball." Lee Trevino has said,

"I haven't seen them all, but I don't know anyone [who] could hit the ball better than Moe Norman" (Macrury, p. 20). Also, in his foreword to Tim O'Connor's biography of Moe, *The Feeling of Greatness,* Trevino wrote, "The simple fact is that when people talk about the great ball-strikers, Moe Norman's name always comes up." Both www.moenormangolf.com and www.moenorman.org have links to videos and more information.

2. "Moe Norman," *Wikipedia,* last modified on 25 September 2011, http://en.wikipedia.org/wiki/Moe_Norman.

3. See www.moenormangolfacademy.org for more information. Thanks to Todd Graves for his help and to John Hamarik for his picture of Moe Norman.

4. Great golfers respected Moe. Tim O'Connor tells of a driving range session in 1971 in which Moe was interrupted by Gary Player to discuss swing mechanics, with the two being joined in short order by Jack Nicklaus and Lee Trevino, all swapping ideas about the perfect swing. (*The Feeling of Greatness,* p. 172). I personally have warm feelings for someone like Moe Norman, who once said, "Golf and math were the only two things that I figured mattered to me in my life" (*The Feeling of Greatness,* p. 25).

5. See Wade, p. 274.

6. See Werner and Greig, *How Golf Clubs Really Work and How to Optimize Their Designs,* p. 21.

7. The deviation is measured from the golfer's average value. That is, the 20-handicapper may have an outside-in swing that averages 5° to the left. In this case, the 3.66° standard deviation would be added to and subtracted from 5° left. The conclusion would be that about two-thirds of this golfer's swings have a swing angle between 1.34° and 8.66° to the left of the target.

8. See Werner and Greig, p. 8. The variables are impact velocity v, angle k of impact (above the horizontal), and the backspin rate ω. The estimate of roll in yards is $0.868v + 0.00173v^2 \cos{(k + 24.4 + 0.0012\omega)}$. For the standard deviations of clubface angle and swing speed, see p. 21.

9. The "basic fade" has swing speed 161 ft/s, driver loft of 12°, swing plane 4.6° to the left, and clubface angle 0.9° to the left.

10. If you speak British English, these would be "weight" and "borrow." A Canadian friend, Richard Grant, once startled me after I hit a

50-foot putt even with the hole but a full 5 feet to the right. Being a kind person, he said, "Nice weight." After briefly thinking that he was complimenting my physique, I did manage the appropriate response: "Too much borrow." But my financial situation is a different story.

11. Werner and Greig, pp. 139–41.

12. Ibid.

13. See B. Hoadley, "How to lower your putting score without improving."

14. The Stimp meter, invented by Edward Stimpson, is a small ramp used to measure the speed of greens. The Stimp number is the number of feet the ball rolls after leaving the ramp. So, high Stimp numbers (11 or higher) indicate fast greens.

15. The actual amount of break is $15 \tan (3.5°) \approx 0.917$ feet.

16. See Dave Pelz, "A study of golfers' abilities to read greens."

17. The curve shown is the trajectory for a ball that starts at $(6, 0)$ ft with an initial velocity of $< -7, -.74 >$ ft/s. The tangent line to the curve at $(6, 0)$ has slope $\frac{-.74}{-7}$, and the y-intercept of the tangent line is $-\frac{.74}{7}$ (6 ft) ≈ 0.634 ft ≈ 7.6 in. This is shown in the figure below.

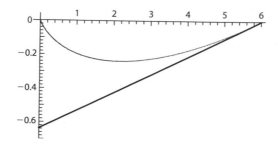

18. See "A study of golfers' abilities to read greens" in *Science and Golf II*.

19. See Dave Pelz, *Putt Like the Pros*, pp. 49–59.

20. Richard Goeres, final project for a class in sport science at Roanoke College.

21. The official record previous to Tiger had been 113 by Byron Nelson, although an article in *Sports Illustrated* claims that the true record is 177 by Ben Hogan. Research found that Hogan had finished in the money in 177 straight tournaments. Although in the early days of the

PGA many tournaments did not have cuts as they are now implemented, the important test for a golfer was finishing high enough to get a paycheck.

22. The curves shown are $y = e^{-(x-70.8)^2/0.98}$ in figure 2.10a, and $y = e^{-(x-70.8)^2/12.5}$ and $y = e^{-(x-68.5)^2/12.5}$ in figure 2.10b.

23. The curves shown in figure 2.11 are $y = e^{-(x-70.8)^2/12.5}$ and $y = e^{-(x-68.5)^2/4.5}$.

Chapter Three. Good Luck Putting

Epigraph: Downs Macrury, *Golfers on Golf*, p. 33.

1. CBS television special, "Fantastic Finishes at the Masters," 2007. As I write, a clip of this, titled "Jack Nicklaus birdies 17th for lead—1986 Masters," is available on YouTube. The photo on p. 44 is one of my favorites. It shows Jack and son/caddie Jackie Nicklaus in a practice round on the 16th tee. In the background is Greg Norman, who seems to be struggling, perhaps anticipating the cruel future ahead of him at Augusta National.

2. In his book *A Feel for the Game*, Ben Crenshaw included this tournament in a short list of golfing events, "Where fate lent a hand," p. 147. In *The Grand Slam*, author Mark Frost indicates that Bob Jones himself probably would have called Nicklaus's victory fate.

3. Dave Pelz, *Putt Like the Pros*, p. 38.

4. See Pelz, *Putt Like the Pros*, pp. 24–28.

5. There is some evidence that putting percentages drop as the round progresses in PGA tournaments. For example, in 2007 from 6 feet away the pros made 73% of their putts in the first hour of play, 71% in the second hour of play, 69% in the third and fourth hours of play, and 66% in the fifth hour of play. From most other distances, the percentages did not drop steadily, and the overall statistics from all distances show only a slight decline as play progresses. Clearly, the pros try to minimize damage to the greens. A quick watering of the greens may also help repair footprints.

6. Werner and Greig, *How Golf Clubs Really Work and How to Optimize Their Designs*, pp. 131–32.

7. Ibid.

8. Yogi Berra includes "90% of short putts don't go in" as one of his Yogi-isms in *The Yogi Book*. In this case, we may want to take seriously the book's wonderful subtitle of *I Really Didn't Say Everything I Said*.

9. See Pelz, pp. 126–30.

10. Werner and Greig, pp. 143–45.

11. The Stimp meter, invented by Edward Stimpson, is a small ramp used to measure the speed of greens. The Stimp number is the number of feet the ball rolls after leaving the ramp. So, high Stimp numbers (11 or higher) indicate fast greens.

12. From *Golf Digest's* "Dialogue on Golf" with Raymond Floyd, February 1994.

13. Kevin Cook, *Titanic Thompson: The Man Who Bet on Everything*, pp. 139–42.

14. Raymond Floyd, *Golf Digest*, February 1994.

15. A full solution of the problem appears in Gualtieri, et al., "Golfer's Dilemma." See also Littlewood and Bollobás, *Littlewood's Miscellany*; and Neimark and Fufaev, *Dynamics of Nonholonomic Systems*, pp. 76–80.

16. Gualtieri, et al., place the origin at the contact point, while Littlewood and Bollobás and Neimark and Fufaev place the origin at the center of the ball. Thanks to Bruce Torrence for help with the Mathematica code for figure 3.5.

17. See Pelz, pp. 31–36. The phrase "lumpy doughnut" is Pelz's.

18. This quote is often attributed to Mark Twain, but German author Kurt Tucholsky has also received credit (from the Germans, naturally). More recently, John Feinstein used it as the title of his book on the PGA Tour.

19. Einstein's paper on Brownian motion was published in 1905, his "annus mirabilis" of unprecedented scholarly production. This paper was important in establishing the atomic nature of life.

20. To be precise, the probability p must be inversely proportional to the distance to the hole, with a maximum probability of q. Then choose b such that $\tan b = 0.02\sqrt{2/q}$. More realistically, b can vary as long as the expected value of $\tan b$ equals $0.02\sqrt{2/q}$.

21. See Pelz, pp. 31–37.

Chapter Four. The Rivalry

Epigraph: Downs Macrury, *Golfers on Golf*, p. 68. Palmer is discussing his final round collapse in the 1966 U.S. Open, losing a 7-shot lead with nine holes to play.

1. Will Blythe's *To Hate Like This Is to Be Happy Forever*, about the college basketball rivalry between the North Carolina Tar Heels and the Duke Blue Devils, is an excellent reflection on the meaning of fierce rivalries. See Tom Stanton, *Ty and the Babe*, for the story of the rivalry of Ty Cobb and Babe Ruth, first in baseball (not only for the title of best in the game, but for control of the style and soul of the game) and then in golf.

2. In *Arnie and Jack*, Ian O'Connor explores the careers of and interactions between Arnold Palmer and Jack Nicklaus, starting with a long-driving contest between young professional Palmer and college freshman Nicklaus. Nicklaus outdrove Palmer, making Arnie mad enough that, in the exhibition match that followed, Palmer broke the course record.

3. See G. H. Hardy, "A Mathematical Theorem about Golf." The rules that I use are exactly as Hardy defined them. Fortunately for me, Hardy contented himself with an approximate calculation for the odds of a player with probability p beating a perfectly consistent player with probability $p = 0$. Hardy, with his characteristic wit, named what I call excellent and poor shots "supershots" and "subshots," respectively. The analysis given here explores the problem further.

One of the book's reviewers suggested a nice alternative way to understand the assumptions. A golfer must accumulate 4 quality points to finish a hole. A P shot is worth 0 points, an N shot is worth 1 point, and an E shot is worth 2 points, so the sequence NEE represents 3 shots with a "wasted" point coming with the second E shot.

4. The exact average (mean) for a par 4 is $4 + p\left[1 - \left(\frac{p}{1-p}\right)^4\right]$. For $p = 0.1$, this computes to $4.099984\cdots$ and for $p = 0.2$ we get $4.1992\cdots$. For par 5s the mean is $5 + p\left[1 + \left(\frac{p}{1-p}\right)^5\right]$, while for par 3s the mean is $3 + p\left[1 + \left(\frac{p}{1-p}\right)^3\right]$. Details are in my April 2010 *Math Horizons* article and companion paper online at www.mathaware.com (and the proceedings publication *Mathematics and Sports*).

5. For example, on the second hole, while my partner was making a 6, I dumped my second into a greenside bunker. Unfortunately, the green it was beside was the 17th. I hit a full 9-iron onto the correct green and dropped a 30-footer for par. Our opponents missed 10-foot and 8-foot

birdie putts, and we halved the hole. I have no idea why they did not strangle us on the spot. Any judge who golfs would have let them off with a warning.

6. Private communication. Chris Conklin did his work as a junior at St. Olaf College, where he was captain of the golf team.

7. See Kevin Cook, *Driven*, pp. 45–46, which gives several revealing case studies of parents supporting their golfing prodigies both well (Ivan Lendl) and poorly. It also gives details of the Scott Robertson Tournament, a top junior tournament in my city of Roanoke.

Chapter Five. Handicap Systems and Other Hustles

Epigraph: Leigh Montville, *The Mysterious Montague*. This book details the amazing golfing feats of the legendary John Montague.

1. The USGA website (www.usga.org) has a number of interesting pages, including a description of the handicap system.

2. A number of articles have been written about the shortcomings of the handicap system. In *The Physics of Golf*, Theodore Jorgensen comes to the conclusion that the better player has the advantage (pp. 105–15). Bingham and Swartz, "Equitable Handicapping in Golf," posit that weaker players are favored in head-to-head matchups.

3. As a college professor, I cringe at using the phrase "A-game" for the 80% mark. About 20% of your scores will be at or better than your handicap, once it is adjusted for slope rating. For a course with a high slope rating, you might *never* shoot only 6 over the course rating.

4. See Francis Scheid, "A General Principle in Golf," pp. 298–304.

5. For more details, see Dean Knuth, "A Two Parameter Golf Course Rating System"; and R. C. Stroud and L. J. Riccio, "Mathematical Underpinnings of the Slope Handicap System."

6. See Stroud and Riccio, "Mathematical Underpinnings," pp. 135–40.

7. The 92% rule holds exactly if the standard deviation is 10% of the mean. As noted later in this chapter, estimates of the actual standard deviations are considerably lower than this, calling the 92% rule into question. The rule is quoted, but not justified, in Stroud and Riccio, "Mathematical Underpinnings."

8. More information can be found at www.ausgolf.com.

9. From Dean Knuth, "History of Handicapping," at www.popeofslope .com.

10. See Richard Stroud, "Proposed Handicap Study."

11. Bingham and Swartz, "Equitable Handicapping in Golf," pp. 170–77.

12. The probability density function (pdf) for a normally distributed random variable with mean μ and standard deviation σ is given by $f(x) = \dfrac{1}{\sigma\sqrt{2\pi}}e^{-(x-\mu)^2/2\sigma^2}$.

13. See F. J. Scheid, "On the Normality and Independence of Golf Scores, with Various Applications," pp. 147–52.

14. Theoretically, all values are possible. To help you visualize the concept of "spread," I am cheating somewhat by focusing on the values of x that produce points with large enough y-values to be visually distinguishable from the x-axis.

15. This again uses the 92% rule from note 7, above.

16. Bingham and Swartz, 170–77.

Chapter Six. The ShotLink Revolution

Epigraph: From www.baseball-almanac.com/quotes/stats5.shtml. Wiles is Director of Research at the Baseball Hall of Fame.

1. In 1941, Ted Williams batted .406, and Joe DiMaggio got hits in 56 straight games. In 1998, Mark McGwire hit 70 home runs, Sammy Sosa had 66, and Ken Griffey, Jr., had 56. In 1968, Bob Gibson had an ERA of 1.12, Denny McLain won 31 games, Don Drysdale pitched $58\frac{2}{3}$ consecutive scoreless innings, and Carl Yastrzemski led the American League with a .301 average. In 1961, Roger Maris had 61 home runs, and teammate Mickey Mantle had 54.

2. Michael O'Keefe and Teri Thompson, *The Card: Collectors, Con Men and the True Story of History's Most Desired Baseball Card*, gives historical background on the popularity of baseball cards.

3. The term "sabermetrics" derives from SABR, the Society for American Baseball Research. Since the B stands for "baseball," the corresponding name for the analysis of golf statistics will have to be different. "Tigermetrics" is my best shot at a catchy name.

4. Many thanks to Mike Vitti of PGA Tour, Inc., for his assistance and guidance. Also, thanks to Steve Evans and Kin Lo for their help.

5. The men's major golf championships, often referred to simply as "the majors," are the Masters Tournament, the U.S. Open Championship,

the Open Championship (commonly known as the British Open), and the PGA Championship.

6. While this quotation appears on many Internet quote sites, such as brainyquote.com and thinkexist.com, I do not know the original source.

7. The classic example is a positive correlation between ice cream sales and murders, "proving" that ice cream causes murder. Although a brain freeze can be uncomfortable, the real reason for the correlation is that both ice cream sales and murders increase when the temperature rises.

8. I used the available online statistics to compute correlations to score for several 2008 PGA tournaments, including majors. In every tournament I looked at, greens in regulation and putts per green in regulation were far and away the best predictors of score.

9. We will see later that missing a fairway costs a player, on the average, about a quarter of a stroke. Thus, it *is* important to hit fairways. The lesson to learn from the correlation study is that the number of fairways hit is not the best predictor of score for the pros. The top ten in a given tournament may comprise a mix of players who hit many fairways and players who hit few fairways. By contrast, the top ten rarely includes a player who misses a relatively large number of greens or takes a large number of putts.

10. You might expect that driving distance would at least correlate with proximity to the hole, which in turn correlates highly with scoring. However, average driving distance for a round has a 0.02 correlation (small and positive!) to average proximity to the hole for the round. For a given (par 4 or par 5) hole, the correlation is -0.09, still quite small. The general idea that longer hitters end up closer to the hole on average is not supported by the data. This statement also needs some context. As we will see, for shots from the fairway, the closer the pros are to the hole, the closer they hit their approach shots. Therefore, extra distance translates to better results, as long as you're keeping the ball in the fairway. Long hitters miss the fairway often enough that distance off the tee does not correlate highly with the quality of the approach shot.

11. The equations can be solved in general using techniques from calculus or linear algebra. The solution can be written simply using vectors and matrices, which are easily implemented on a computer.

Chapter Seven. Lags and Gags

Epigraph: From the NBC telecast, June 15, 2008.

1. See Rocco Mediate, *Are You Kidding Me?*

2. Outside of 20 feet, the match between the function and the data is quite poor. The function approaches 0 far faster than does the percentage of putts made.

3. See Bob Rotella's "How to Drain Them Like Jack."

4. As a sample calculation of significance, if there was no difference (other than randomness) between made par and made birdie percentages, then we would expect that 99.2% of the 9,150 birdie putts from inside 3 feet—9,077 putts—would be made. Actually, only 8,958 of these birdie putts were made. The standard deviation of a binomial distribution with $n = 9150$ and $p = 0.992$ is approximately 8.5, so the observed value of 8,958 is a full 14 standard deviations away from the expected value *if* the averages were the same.

5. Thanks to Chip Sullivan for his time and help at Hanging Rock Golf Club, as he prepared for the 2010 PGA Championship at Whistling Straits.

Chapter Eight. Chips and Flops

Epigraph: From "10K Truth Quotes on Golf," www.10ktruth.com/the_ quotes/golf.htm. The rest of the quote is: "Fate has nothing to do with success or failure, because that is a negative philosophy that indicts one's confidence, and I'll have no part of it."

1. The 1987 Masters ended in a three-way tie between Greg Norman, Larry Mize, and Seve Ballesteros. Seve dropped out with a bogey on the first playoff hole, number 10 at Augusta. On the par 4 eleventh, Mize left his second shot well to the right of the green. Norman hit a conservative shot to the front of the green, some 30 feet from the hole. Mize played a bump-and-run that took two hops in the fringe before bouncing onto the green. The chip had a fair amount of speed (Mize estimates it would have gone 4 to 8 feet past the hole) when it hit the pin and dropped.

2. We can derive an equation of the most visible of the parabolas in figure 8.8. Start with a point for which $x = 0$ and $y = H > 0$. Suppose the real value of B changes very slightly so that it rounds down instead of up. In effect, we have a sudden change from a reported value of B to a reported value of $B - 1$. By our formula for y, the new y-value is $\tilde{y} = y - \frac{2B-1}{2d}$.

suming that B and d are both much larger than 1, then the new y-value is approximately $\tilde{y} = y - 1$. The new x-value will be obtained from the equation $x^2 = H^2 - \tilde{y}^2 = y^2 - \tilde{y}^2$. Substitute the value for $\tilde{y}$ found above, square it, and simplify to get $x^2 = \frac{2B-1}{d}y - \left(\frac{2B-1}{2d}\right)^2$. Solving for y gives $y = \frac{d}{2B-1}x^2 + \frac{2B-1}{4d}$. Assuming that B and d are much larger than 1 and that they are approximately the same (for example, a shot from 300 feet out will travel approximately 300 feet), then $y \approx \frac{1}{2}x^2 + \frac{1}{2}$, which (when converted to feet, giving $y = 6x^2 + \frac{1}{24}$) matches the innermost parabola nicely. A change from a value of d to $d+1$ produces the same approximate equation.

3. Rounding off is at the center of a long-standing controversy over how to apportion the United States House of Representatives. The first presidential veto in U.S. history was George Washington's veto of Thomas Jefferson's method for apportioning representation in the House in favor of a different method of rounding. Mathematically, all methods of rounding produce undesirable paradoxes in the apportionment of the House. For example, increasing the size of the House can result in a state losing a representative (known as the Oklahoma paradox). The general effect seen in figure 8.8 is related to the butterfly effect. This refers to the sensitivity and unpredictability of the weather, in that a small change (the air disturbed by a butterfly flapping its wings in Brazil) can have a large effect (the formation of a tornado in Texas).

Chapter Nine. Iron Byron

Epigraph: From Thinkexist.com. Hogan also chided an overly exuberant Lanny Wadkins with the warning, "I don't play jolly golf."

1. Screenplay by Mark Frost, based on his book of the same title.

2. See Mark Frost, *The Match*, pp. 2–4.

3. See Tom Wishon, *The Search for the Perfect Golf Club*, pp. 7–9. Wishon calls this the "dreaded vanishing loft disease."

4. The equations are as in note 10 of chapter 1, with $x(0) = y(0) = 0$, $x'(0) = v \cos(\theta)$, $y'(0) = v \sin(\theta)$, $\omega(0) = \omega$. For a given club loft angle A and swing speed S, the launch angle of the ball is approximated by

$$\theta = \tan^{-1}\left(\frac{1.67 \tan A}{0.5 \tan^2 A + 1.83}\right),$$

and the launch speed of the ball is given by

$$v = S \frac{1.83 \cos A}{\cos (A - \theta) + \frac{46}{180} \cos A \cos \theta},$$

with an initial spin rate of

$$\omega = [7.143S] \sin A.$$

The adjustments for club length were based on a half-inch difference in lengths of clubs. Clubhead speeds were $S = 132$ ft/s for the 5-iron, $S = 130.3$ ft/s for the 6-iron, $S = 128.6$ ft/s for the 7-iron, $S = 126.9$ ft/s for the 8-iron, and $S = 125.1$ ft/s for the 9-iron. See Theodore Jorgensen, *The Physics of Golf*, pp. 123–31.

5. The top tens for 2008 for average approach distances from the rough (in ft) are given Appendix A, table A9.1.

6. The best fit quadratic function is $0.049x^2 + 0.171x + 2.56$. The fit is not very good for the first few data points, but beyond 50 yards, the fit is excellent. The best fit quadratic for the standard deviations of the amount offline is $0.035x^2 + 0.178x + 3.261$.

7. The standard deviations of the amounts short and long increase as the distance increases, but in more of a linear than quadratic fashion.

Chapter Ten. The Big Dog

Epigraph: From Downs Macrury, *Golfers on Golf*, p. 59. My favorite golf moment is a mid- or long-iron hanging in the air, with time seeming to stop while my eyes go down to the pin and up to the ball to verify that the ball is heading right at the pin.

1. However, you do get a free drop if your ball lands in the sheep droppings that litter the fairways.

2. From Paul DiPerna and Vikki Keller, *Oakhurst: The Birth and Rebirth of America's First Golf Course*, p. 156. Many thanks to Nancy Midkiff for her tour of the Oakhurst museum. She cleared the way for me to play the course when it was officially closed to the public, took excellent care of me and my friend John Selby, and mailed me an autographed copy of the book. She and Mr. Keller made us feel very welcome.

3. Unfortunately, the only rhythm I developed was a waltz tempo of backswing–downswing–"Fore, right!" Mr. Keller and Ms. Midkiff had

warned me that my fast swing was not likely to work well with the head-heavy hickory driver.

4. See John Andrisani, *The Bobby Jones Way*, p. 32.

5. On a recent visit to Golf Mart (many thanks to Randy Agee), I hit several shots with my 10-year-old driver and a variety of new drivers. Each of the modern drivers had a clubhead about twice the size of my "midsize" driver. Clubhead speed (CS), ball speed, launch angle, spin rate, and estimated carry distance for representative swings of five different clubs are shown below.

Driver	CS	Speed	Angle	Spin	Carry
	(mph)	(mph)	(°)	(rpm)	(yd)
A	101.6	150.4	10.5	1766	239.9
B	100.1	146.5	13.6	2827	267.7
C	102.8	147.7	11.9	3913	259.2
D	104.6	150.8	12.5	3786	269.0
E	100.8	147.2	9.6	3121	245.7

Notice that while the clubhead speeds stayed fairly constant, all of the other variables changed dramatically. Unfortunately for my bank account, my driver (club A) did not fare well. Driver B rated the highest. Although the carry distance for driver D was larger, the lower spin rate of driver B would likely produce more roll and, therefore, the largest overall distance (in spite of driver B having the lowest clubhead and ball speeds).

6. See Tom Wishon, *The Search for the Perfect Golf Club*, pp. 13–18.

7. There are many good stories about the longest shot ever hit, including hitting drives into trucks or down roads, as in the movie *Tin Cup*. The official record for longest drive in a tournament is 515 yards by the remarkable Mike Austin, who was 64 years old at the time. Other official records tend to involve teeing off on the runway at an airport. However, without a doubt the longest drive ever hit was by cosmonaut Mikhail Tyurin on November 22, 2006. Tyurin stroked a 6-iron into space while strapped to the International Space Station. NASA submitted a low estimate for the shot (which they described as partially shanked) of slightly over a million miles. Russian estimates top a billion miles.

8. Not a Cinderella story in the *Caddyshack* fantasy sense. Entering the PGA Championship that year, John Daly was an unknown rookie

struggling to keep his card. He was the ninth alternate, such a longshot to even get into the tournament that he did not play a practice round at the course, Crooked Stick (near Indianapolis). He shot 69 and 67 in the first two rounds and calmly outplayed the field over the weekend to win. A little-known fact is that a spectator, Tom Weaver, had died from a lightning strike in the first round at Crooked Stick. After his win, Daly quietly donated $30,000 of his $230,000 first prize to the Weaver family. Ironically, Daly's career has followed a crooked stick, with highly publicized successes and failures on and off the course. He remains one of the most popular players in the game.

9. Both quotes are from Macrury, *Golfers on Golf*, pp. 21 and 12, respectively.

10. Mike Vitti, PGA Tour, personal communication.

Chapter Eleven. Tigermetrics

Epigraph: See Don Wade, *Talking on Tour*, p. 368.

1. See Wade, *Talking on Tour*, p. 164. Later, Hogan was asked if he wanted to go to the driving range to watch Faldo hit balls. Hogan inquired as to whether Faldo used Hogan golf clubs. When told that Faldo did not, Hogan responded, "Then I think I'll just sit here and finish my wine."

2. As evidence that 27 is not enough putts, note that Tiger placed 17th out of 200 on putts made between 8 and 9 feet. It is unlikely that anybody is woefully deficient from 8 feet but highly accurate from 9 feet.

3. Each hole is counted only once. If the first putt is from 60 feet, then I compare the number of putts to the Tour average from 60 feet. I do not consider how long the second putt was. There is no distinction made between great lag putts and poor lags with long makes for the 2-putt. What is being measured is total putting performance. Also, instead of using the actual average number of putts for a given year, I use a number derived from the fitted curve to the number of putts. For ratings of other skills, I also use fitted curves in the computations.

4. There is some evidence along these lines for 2007. In the tournaments that Tiger played in, the average number of 3-putts was 0.623, and the average number of total putts was 29.1. In the tournaments that Tiger did not play in, those averages dropped to 0.609 and 28.6, respectively. There is more on this topic at the end of the chapter.

5. The number of shots of different types is not arbitrary. In 2008, 28,349 shots were taken from the fairway from 4–50 yards, 125,130 shots from the fairway from 50–200 yards, 17,063 shots from the fairway from 200–250 yards, 35,248 shots from the rough from 4–50 yards, and 39,600 shots from the rough from 50–200 yards. The weights are roughly proportional to the number of shots in each category.

6. From an interview with Laura Hill, PGATOUR.com.

7. See Douglas Fearing, Jason Acimovic, and Stephen Graves, "How to Catch a Tiger: Understanding Putting Performance on the PGA Tour," pp. 1–49.

Chapter Twelve. More Rating Systems and Tiger Tales

Epigraph: From Downs Macrury, *Golfers on Golf*, p. 61.

1. See my UMAP Module 725 in "A Mathematical Rating System" or my website, www.roanoke.edu/staff/minton/bynumbers.html. More details are given in the "Back Tee" section of chapter 12.

2. This large point differential is contrasted with NASCAR's point system in Scott Berry, "Is Second Place the First Loser?"

3. The actual graph shown in figure 12.1 is that of $.000039(x - 35)^4 +$ $.00009(x - 35)^2$, whereas the aging function found by Berry et al. is empirical and does not have a simple formula. The actual aging function graphed in Berry et al. does not appear to be symmetric about $x = 35$, although I tried to match the shape of their graph reasonably well.

4. Private communication with Scott Berry.

5. See John Zumerchik, *Newton on the Tee*, pp. 197–206.

6. See Jennifer Brown, "Quitters Never Win," pp. 1–35.

7. The interpretation given by some in the media, that Tiger is so intimidating that he starts off each tournament with a de facto 2-shot lead, is not warranted. That might be what is happening, but Brown's work was focused on the overall effect on the players when Tiger is winning every tournament in sight. Rich Beem, Zach Johnson, Trevor Immelman, Rocco Mediate, Y. E. Yang, and others have stood up nicely to Tiger in major championships.

Glossary

Airmail a green. Hit a ball too far, so that it flies over the green.

Approach shot. A shot from several yards off the green which is expected to reach the green. Approach distance is the distance to the hole after the shot.

Asymptote. A line or other curve to which a given curve becomes infinitely close.

Back nine. The last nine holes on a golf course. This is preceded by the "front nine" to make a round of 18 holes.

Backspin. The most common spin on a golf shot. Viewed from behind, a ball with backspin rotates such that the top of the ball moves directly toward the golfer. The spin vector is to the right.

Back tees. On a hole with several locations for tee shots, the back tee is the tee location that is farthest from the hole. Most courses have back tees (also called "the tips"), white tees (for average golfers), gold tees (for seniors), and red tees (for women).

Bag. The golf bag that holds the golfer's 14 clubs, balls, and accessories used during a round. If you're Rodney Dangerfield in *Caddyshack*, it also holds a keg.

Ball speed. The initial speed of the ball, immediately after contact.

Bernoulli trial. An experiment in which the random outcome can be one of two values, such as a made putt or missed putt. In a sequence of Bernoulli trials, the outcomes are independent and identically distributed (e.g., the probability of making each 5-foot putt is the same and does not depend on whether other 5-foot putts have been made or missed).

Best ball. A team competition in which the team score on a hole for each two-player team is the better of the scores of the individuals on the team. Usually, match play determines the winner.

Best-fit line. A line that comes closest to fitting the points in a data set, where "closest" usually means that the sum of the squares of the errors is minimized.

Bethpage Black. A public course in New York that has hosted the 2002 and 2009 U.S. Opens. It is long and tough, and proud of it. Highly visible signs are posted reading, "The Black Course is an extremely difficult course which we recommend only for highly skilled golfers."

Birdie. A score on a hole which is one stroke less (better) than par (e.g., a 3 on a par-4 hole).

Birdie putt. A putt which, if made, would result in a birdie.

Bogey. A score on a hole which is one stroke more (worse) than par (e.g., a 5 on a par-4 hole). Originally, bogey and par referred to the same score, which is the score that would be achieved by an unbeatable "Bogey man." By 1900 or so, improvements in equipment had made the old bogey standard obsolete for good golfers, and a new "par" was set.

Break of putt. From the golfer's perspective, the left/right curving of a putt. An important part of reading a putt is determining how much break to play—that is, how far to the left or right of the hole to aim so that the ball breaks into the hole.

British Open. One of the four major championships in men's professional golf. Played on a variety of courses in England and Scotland, it is the oldest of the major championships. The official name of the tournament is The Open Championship.

Brownian motion. A mathematical process that models the motion of a light particle being constantly bombarded by random forces.

Bump-and-run. A chip shot that is "bumped" a short distance in the air, lands short of the green, and rolls (runs) a long distance to the hole.

Bunker. A cavity in the ground, usually filled with sand (aka "sand trap").

Caddie. A person who carries a player's bag, and often gives advice. The word derives from the French *cadet*, or boy. Mary Stuart, Queen of Scotland and an avid golfer, is given credit for applying the name to golf.

Cane. A score of 7 on a hole.

Captain's choice. A popular format for casual tournaments. On each hole, each player in the group (usually a foursome) hits a tee shot, the group captain chooses the best tee shot, all other balls are picked up, and everybody hits from the location of the best shot. The best of the second shots is chosen, and the process continues until the ball is holed out. Also called a "scramble."

Carry. The distance that a ball travels from starting point to initial landing point (first bounce).

Chip. One type of shot from just off the green, usually played with a shor
swing, in which the ball pops into the air and then rolls along the green
toward the hole.

Club difference. The change in number of iron used due to adjustments to
terrain, wind, or other conditions. For example, if a golfer is at a distance
from the hole where a 7-iron would normally be the correct club choice,
a 2-club difference due to the green being at the top of a hill would make
a 5-iron the correct choice. Since golfers typically have a 10-yard gap
between irons, other ways to express this 2-club difference are "take two
clubs more" and "the shot plays 20 yards longer."

COR. Coefficient of restitution, a measure of the liveliness of a ball. Officially,
COR equals the speed of the ball immediately after contact divided by the
speed of the ball immediately before contact. If no energy is lost, COR = 1.
If the ball is mush and loses all of its speed, COR = 0.

Correlation. A statistic that measures the extent to which two sets of data
can be related by a first degree polynomial (line). A correlation of 1 or –1
means that one variable can be exactly obtained by substituting the other
variable into an appropriate equation of the form $y = mx + b$.

Course rating. The score that a scratch golfer should average on a course.
Course ratings are used to determine handicaps.

Cut. The process by which the large group of golfers that play the first few
rounds of a tournament is reduced to a smaller group to play the last
round(s) of the tournament. A typical PGA tournament starts with 132,
144, or 156 players for the first two rounds. The field is then cut to the top
70 golfers plus ties for the last two rounds.

Dimples. The indentations on a golf ball. Dimple patterns are symmetric
but vary from ball to ball. (Some illegal balls have been designed with
asymmetric dimple patterns, which cause the ball to rotate around a
specified axis and can reduce hooks and slices.) The number of dimples
also varies from ball to ball, although 336 is a common number.

Double bogey. A score on a hole that is two strokes more (worse) than par
(e.g., a 6 on a par-4 hole). The amount over par is twice that of a bogey. If
you "double" a hole, you make a double bogey.

Double eagle. A score on a hole that is three strokes less (better) than par
(e.g., a 2 on a par-5 hole). This is also called an "albatross" (bigger than
an eagle). Reportedly, Gene Sarazen referred to his famous 1935 double
eagle at the Masters as a "dodo," a very rare bird.

wnhill shot. A shot that travels from a starting point that is at a higher elevation than the landing point. A downhill *lie* indicates that the ground near the ball slopes downhill as you move a few feet toward the hole.

Drag force. A force that opposes the motion of a moving object, with a direction that is exactly opposite that of the object's motion and depends on the speed of the object.

Draw. A long shot that curves slightly in a hook direction (right-to-left for right-handed golfers, left-to-right for left-handed golfers).

Drive. A tee shot hit with a driver. Sometimes the term refers to a long tee shot hit with a club other than a driver.

Driver. The longest and biggest club, used to hit the longest shots. The clubheads of drivers are generally large (460 cc) and are now molded in a variety of shapes. The driver is one of several clubs referred to as "woods," an anachronistic term referring to the fact that the clubheads were made primarily of wood until the 1980s.

Driving range. A large, wide-open area for players to hit practice shots of all types.

Eagle. A score on a hole that is two strokes less (better) than par (e.g., a 2 on a par-4 hole). An eagle is a large bird, so we were spared the name "double birdie."

Equilibrium value. A value of a variable in an equation which represents a constant solution; if that value is ever reached, the variable retains that value, as the forces that would change the value are balanced.

Expected value. The mean (average value) of a random variable, so that in some sense this is the value that you would expect to see.

Face angle. As viewed from above, the angle (in a horizontal plane) between a line from the ball to the target and a line from the ball that is perpendicular to the club face at impact. A nonzero angle typically results in a shot that is offline and/or curves.

Fade. A long shot that curves slightly in a slice direction (left-to-right for right-handed golfers, right-to-left for left-handed golfers).

Fairway. A stretch of well-mowed grass, usually connecting the tee area to the green, which is a preferred landing area for shots that are not expected to reach the green. In the fairway, you usually have a good lie for your next shot.

Fairways hit. The number of times a tee shot, excluding par-3 holes, finishes in the fairway (as opposed to the rough). The same term is used as shorthand for the fraction or percentage of times the golfer hits the fairway.

FedEx Cup. A year-long, points-based competition on the PGA Tour. The FedEx Cup playoffs consist of four tournaments, culminating in the Tour Championship.

Firm putt. A putt which is still moving at a high speed when it reaches the hole. If the putt is too firm, it might lip out.

Flop shot. A short shot taken from near the green with a very long swing and lofted club. The ball goes almost straight up in the air and lands with very little horizontal speed.

Fringe. A small area around the green that is less closely mowed than the green but more closely mowed than the fairway.

Front nine. The first nine holes on a golf course. This is followed by the "back nine" to make a round of 18 holes.

Frozen rope. A shot whose trajectory is so straight (and usually close to the ground) that it resembles a rigid rope.

Gallery. The group of spectators watching a tournament.

Game theory. A mathematical field in which "games" of competition are analyzed to determine optimal strategies. Generally, each player in the game has a set of strategies from which to choose. Each player simultaneously chooses a strategy, and the set of choices determines the outcome (payoff) of that round of the game.

Go low. Shoot a low (good) score for a round.

Going for the green. Trying to reach the green with a shot. When faced with a long shot over water, the golfer must decide whether to go for the green (try to hit over the water) or lay up (hit a shot that will stop before reaching the water).

Green. A closely mowed portion of the course around each pin (hole). On a green, you normally putt the ball and let it roll toward the hole.

Greens hit in regulation. The number of times, during a round, that a golfer reaches a green in the regulation number of shots. The regulation number, always two lower than the par on the hole, is the number of times a good golfer should hit the ball to get it on the green.

Gutta-percha ball. A ball made of gutta-percha, an inelastic latex produced from the sap of gutta-percha trees. Gutta-percha balls were used in the late 1800s and early 1900s and typically were scored to provide some of the aerodynamic benefits given by dimples. The more modern balata ball is made from a material that is very similar to gutta-percha.

Halve a hole. To tie a hole by making the same score as your opponent.

Handicap. A number assigned to a golfer to indicate how many strokes worse the golfer is than a scratch golfer. A golfer who has a handicap of 10 (a 10-handicapper) and shoots a 78 would have a net 68, which could be (more or less) fairly compared to the net score of some other golfer to determine the winner of a bet or a tournament.

Hazard. A feature of a golf course, typically to be avoided, in which some actions of a golfer are restricted (e.g., the golfer cannot ground the club— let the club touch the ground before the swing). Common hazards are water hazards and bunkers (sand traps).

Hickory sticks. An early style of golf club (becoming obsolete in the 1930s) in which the shaft of the club is made from hickory wood. Hickory shafts gave way to steel shafts, and graphite and other composite materials are used today.

High side. From the golfer's perspective, the half of the green (either to the left of the hole or to the right) that has higher elevation near the hole. To make a putt, the golfer aims some amount to the high side and lets the putt break down to the hole. If you miss a putt on the high side, you played too much break (aimed too far away from the hole) for the speed of the putt.

Hole. (a) The cylindrical hole in the ground into which the ball must eventually be hit. (b) One of 18 layouts on a course, starting on a tee box and ending on a green. (c) A verb meaning that a shot goes into the hole; when you "hole out," your ball goes into the hole and you have finished the hole.

Hook. A long shot that curves significantly from right-to-left for right-handed golfers or left-to-right for left-handed golfers.

Hybrid. A type of club that is designed as a cross between traditional woods and irons.

Intermediate rough. A strip of grass parallel to the fairway which is mowed lower than the rough but not as low as the fairway.

Iron. One of several numbered clubs with a thin, grooved face. The numbers relate to the length of the shaft and the loft of the face. The higher the number, the shorter the shaft and the more lofted the face; both differences cause shots with high-numbered irons (called short irons) t travel less horizontal distance than shots with low-numbered irons (called long irons). Irons range from a 1-iron (mostly obsolete) to a 9-iron and various wedges.

Iron shot. A shot using an iron.

Jab. A putting stroke that is an abrupt stab at the ball instead of a smooth backswing and long follow-through.

Lag putt. A putt, usually from a relatively long distance, in which the primary goal is to have the ball stop close to the hole. This usage is similar to that used in pool.

Laying up. A strategy in which the golfer does not go for maximum distance but chooses to hit the ball to an advantageous location. Instead of trying to hit a long shot over water, you might choose to lay up to a spot just short of the water.

Lie. The quality of the ball's position on the ground. If you have a good lie, the ball is sitting up nicely on a smooth, flat patch of well-mowed grass. A bad lie can result from the surrounding grass being too long, or from there not being any grass around, or from the surrounding grass being sloped significantly in some direction, or other irregularities in the way the ball is positioned.

Lift force. A more common, less precise term that can refer to a Magnus force, usually used when the Magnus force has an upward component.

Line of putt. The direction in which a putt should start to enable it to go into the hole.

Lip. The edge of the hole.

Lip-out. A putt in which the ball hits a portion of the hole but catches the edge and bounces or spins away from the hole.

Loft of club. The angle between the plane of the clubface and a plane perpendicular to the ground. Common lofts range from 7° or 8° for some drivers to 60° or more for lob wedges.

Long iron. An iron that has a long shaft and is used to hit a long distance; 3-, 4-, and 5-irons are usually considered long irons.

LPGA. The Ladies Professional Golf Association.

Lumpy green. A green that looks smooth but has imperceptible indentations where people have stepped and the grass has not sprung back completely. These "footprints" can change the path of a putt.

...us force. The force resulting from the rotation of a moving object, which causes an asymmetry in the flow of air around the object. If the spinning object causes the air to deflect in one direction, there is an equal and opposite force that deflects the ball, causing its path to curve.

Make zone. An image showing the range of values of some variables corresponding to a made putt. One type of make zone shows the stopping points for made putts (if the hole had not been there). A different type of make zone shows combinations of initial speed and initial line for made putts.

Making the cut. Playing well enough in the first few rounds of a tournament to qualify for the last round(s) of the tournament.

Masters. One of the four major championships in men's professional golf. It is always played at Augusta National Golf Club in Georgia the first week of April. Started by the great Bob Jones, it is a "tradition unlike any other."

Match play. A 2-person or 2-team competition in which each hole is a separate contest. The winner of match play is the person or team that wins the most holes. The result of a match is described in terms of when the match ended. For example, if one person leads by four holes (has won four more holes than the other person) and there are only three holes left, then there is no need to play any further, and the winner takes the match 4 and 3.

Mean. One of several statistical averages of a collection of data or random variable. If there is a finite number of data points, the mean is the familiar sum of the values divided by the number of data points.

Medalist. The person with the best score in a one-round tournament. For this reason, stroke play is sometimes called "medal play."

Missed green. A situation where the golfer is not on the green after taking the regulation number of strokes for that hole.

Modeling. A mathematical construction that is intended to capture in mathematical terms the most important aspects of some real-world situation.

MOI. Moment of inertia, a measure of the resistance to rotation for a particular object about a particular axis. The higher the MOI, the more resistance there is to changes in rotation rate. A club with a high MOI for the appropriate axis will rotate (twist) less for an off-center hit.

19th hole. Since a round consists of 18 holes, the 19th hole is a "watering hole" to sit down, talk, eat, and drink.

Off-line. A shot that does not finish on the line between the golfer and the hole.

One more club. A change of (typically) irons to a number that is one smaller. For example, if you cannot hit an 8-iron far enough to reach the green, you might need to take "one more club" (a 7-iron) to get the extra distance needed.

One-putt. Making your first putt on a hole.

Open clubface. Having the clubface "point" off-line in a particular way; that is, the line perpendicular to the clubface does not lie in the vertical plane containing the golfer and the target. For a right-handed golfer, an open clubface points to the right of the target (and a closed clubface points to the left of the target). For a left-handed golfer, an open clubface points to the left of the target (and a closed clubface points to the right of the target).

Par. A standard score for a hole. Par consists of the regulation number of shots, the number a good golfer should take to get to the green, plus 2 putts. For a professional golfer on an easy course, matching par is not good, and a round that is average ("par for the course") might be several strokes under par. For a casual golfer, par is an exacting standard that is only occasionally met, and an average round is much worse than par.

Par save. A situation in which the golfer does not hit the green in regulation (misses the green) but makes a par nonetheless. It can also refer to a lengthy putt made to save par.

pdf. A probability density function. A function that contains all information about the probabilities of a random variable assuming different values. For a continuous pdf, the area under the curve from a to b is the probability that the variable assumes a value between a and b.

Penalty stroke. Under some circumstances, the golfer must add a stroke (the penalty stroke) as part of the process of resuming play on a hole. After hitting a ball into water, the golfer usually drops a new ball in a location designated by rule and also adds a stroke to the running score. For a lost ball or a ball hit out of bounds, you lose "stroke and distance," meaning that you add a penalty stroke and hit the next shot from the original location.

Perimeter weighting. A style of clubmaking in which extra weight is manufactured into the edge of the clubface of an iron, resulting in an increased MOI.

PGA Championship. One of the four major championships in men's professional golf. Played on a variety of courses in the United States, it is run by the Professional Golfers of America (which is different than the PGA Tour).

PGA Tour. The organization that runs most of the events played in the United States. It is distinct from the PGA and the USGA. Most of the golfers you watch in tournaments belong to the PGA Tour, whereas the pro at your local course probably belongs to the PGA.

Pin. Sometimes synonymous with "hole," the pin is the pole stuck into the hole, usually with a flag at the top. A shot that is online is going "at the hole" or "at the pin" or "at the flag."

Pitch. One type of shot from just off the green, usually played with a highly lofted club, in which the ball is hit high into the air and has a small amount of roll after landing on the green.

Pitch out. A short shot hit when the golfer is unable to make good progress to the hole, due to a tree in the way or high rough or other circumstance. Even though it feels like wasting a stroke, you pitch out ("take your medicine") to a location from which you can hit a good shot.

Polar coordinates. A mathematical system of describing two-dimensional points by measuring the distance from the point to the "origin" (a special fixed point) and the angle counterclockwise from the positive x-axis to the line segment connecting the point and origin.

Pro-Am day. A day, usually the day before a tournament starts, in which the pros play with amateur partners of varying ability.

Pull. A shot which starts off heading to the left of the target for a right-hander or to the right of the target for a left-hander.

Push. A shot which starts off heading to the right of the target for a right-hander or to the left of the target for a left-hander.

Putt. For the purpose of official statistics, a putt is a shot taken from on the green. Typically, a putter is used to roll the ball along the green. A shot from off the green using a putter is not considered a putt.

p-value. When referring to a simulation of a golf round using G. H. Hardy's rules, p is the probability of a given shot being excellent and also the probability of that shot being bad.

Q-school. The PGA Tour Qualifying Tournament. The top 125 money winners from one season automatically qualify as regulars for the following season. A multi-stage qualification process for players outside the top 125 culminates in a 6-round tournament in which approximately 150 players compete for 25 spots as regulars on the PGA Tour. These players are said to "earn their PGA card."

Random variable. A mathematical function whose values are subject to randomness, although the distribution of values may be known.

Random walk. A mathematical sequence in which the next value is either one more or one less than the previous value, each with probability one-half. A physical realization of a random walk is a person who with each step wobbles randomly to the left or the right.

Range finder. A device that computes the distance from a golfer to a given feature, such as a pin or water hazard.

Read the green. To determine how the ball is going to roll on a green, specifically to know how hard and in what direction to hit a putt.

Regression to the mean. A statistical phenomenon in which extreme values of a random variable are followed by values that are closer to the mean.

Rough. The portion of a golf hole to either side of the fairway which consists of long grass and other vegetation that is not golfer-friendly. In the ShotLink system, "rough" is a broader designation meaning "not in the fairway" and includes trees and hardpan (hard, compacted ground).

Round of golf. Playing the 18 holes on a golf course.

Ryder Cup. A semi-annual competition between the United States and Europe, featuring hotly contested matches between many of the world's best players in best ball, alternate shot, and match play singles formats.

Saving par. Making a par in spite of having missed the green or making a long putt for par.

Scrambling. In general, the expression refers to hitting good recovery shots such as bunker shots, chips from off the green, shots between and around trees, or long putts. The official statistic equals the fraction of time that a player misses a green and makes par or better.

Scratch golfer. A golfer with a 0 handicap. The origin of the term is related to the phrase "made from scratch" which indicates no head start or handicap.

Short. A shot which, from the golfer's perspective, does not go as far as the hole.

Short iron. An iron for hitting short shots, short in length but with a high number, such as a 9-iron or wedge.

ShotLink. A system of lasers and volunteers run by the PGA Tour to collect data on the location and outcome of every shot in every round.

Side door. From the golfer's perspective, the left or right edge of the hole.

hill putt. A putt which will not, due to the slope of the green, roll directly toward the hole.

kins. A multiplayer competition in which the golfer with the best score on a hole wins the bet (the "skin").

Slice. A long shot that curves significantly from left-to-right for right-handed golfers or right-to-left for left-handed golfers.

Slope rating. A number assigned to each course to indicate how hard it plays for non-expert golfers and used in the USGA handicap system. An average slope rating is 113, and higher ratings indicate courses that play significantly harder than average for casual golfers.

Slow green. A green on which a rolling ball stops quickly, due to the length or the type of the grass. A slow putt can be due to a slow green, putting against the grain of the grass, or putting uphill.

Snowman. A score of 8 on a hole.

Spike marks. Golf shoes have spikes of different types to keep the golfer's feet from slipping during the swing. Sometimes these spikes, especially pointed metal spikes, pull up portions of dirt or grass as they exit the ground. Spike marks can deflect a putt significantly but by rule may not be smoothed down by a golfer who is about to putt.

Spin the ball. To impart a lot of spin to the ball on a shot. With enough spin, the ball will not bounce or roll forward very much after landing.

Standard deviation. A statistical measure of the magnitude of the differences among values of a data set or random variable; in other words, it measures the dispersion or spread of values from the mean.

Standard error. A statistical estimate of the standard deviation based on a set of measurements.

Stimp meter. A ramp used to measure the speed of a green. The Stimp reading is how far in feet a ball rolls after leaving the ramp.

Stroke play. A type of competition in which the player with the fewest total strokes for the round wins.

Stymie rule. In singles match play until 1952, balls were not marked on the green unless they were within six inches of each other. If an opponent's ball was in between your ball and the hole, you were "stymied" and could either try to putt around the ball or try to chip over it.

Swing angle. As viewed from above, the angle (in a horizontal plane) between a line from the ball to the target and the (tangent) line of the clubhead's path near impact. A nonzero angle typically results in a shot that is off-line and / or curves.

Swing plane. The plane containing the rotating lines from clubshaft to club-head near contact. For most golfers, there is a different plane formed by the backswing. The intersection of the swing plane and a horizontal plane gives the line used to measure swing angle.

Swing speed. The speed of the clubhead at impact. This has been measured to be about twice the speed of the golfer's hands at impact. Casual golfers have swing speeds in the 75- to 95-mph range, pros are in the 115- to 130-mph range, and long drive champions regularly reach 150 mph.

Tee shot. The first shot on a hole, usually with the ball sitting on a tee.

Three-putt. Needing three putts to get the ball in the stupid hole.

Trajectory. The path followed by a struck ball.

U.S. Open. One of the four major championships in men's professional golf. Played on a variety of courses in the United States, it is run by the United States Golf Association.

Under par. Having a score that is less (better) than par. A golfer who is currently "3 under" (written –3) has taken 3 fewer strokes than the pars for the holes that have been played.

Up-and-down. From a location off the green, needing only one shot (up on the green) plus one putt (down into the hole) to finish the hole.

Uphill shot. A shot that travels from a starting point that is at a lower elevation than the landing point. An uphill lie indicates that the ground near the ball slopes uphill as you move a few feet toward the hole.

Vector. A mathematical entity that has both size (magnitude) and direction.

Wedge. One of several clubs that have large lofts and are used to hit high, short shots. Common wedges are pitching wedges (loft around 48°), gap wedges (about 52°), sand wedges (about 56°, with special "bounce" properties to help hit good shots from sand), and lob wedges (about 60°).

White tees. On a hole with several locations for tee shots, the white tee is the tee location that is used by average golfers. On some holes, there can be 50–100 yards' difference between the white tees and the back tees.

z-score. Also called a standard score, a z-score estimates the number of standard deviations that a value of a random variable is above or below its mean.

References

Articles

Beasley, D., and T. Camp. "Effects of Dimple Design on the Aerodynamic Performance of a Golf Ball." *Science and Golf IV*, 2002.

Berry, Scott. "Drive for Show and Putt for Dough." *Chance* 12, no. 4 (1999), pp. 50–55.

Berry, Scott. "How Ferocious is Tiger?" *Chance* 14, no. 3 (2001), pp. 51–56.

Berry, Scott. "Is Second Place the First Loser?" *Chance* 18, no. 1 (2005), pp. 55–59.

Berry, Scott, Shane Reese, and Patrick Larkey. "Bridging Different Eras in Sports." *Journal of the American Statistical Association* 94, no. 447 (Sept. 1999), pp. 661–76.

Bingham, Derek, and Tim Swartz. "Equitable Handicapping in Golf." *American Statistician*, 54, no. 3 (Aug. 2000), pp. 170–77.

Brown, Jennifer. "Quitters Never Win: The (Adverse) Incentive Effects of Competing with Superstars." Job Market Paper, University of California, Berkeley, Nov. 2007.

Carnahan, J. V. "Experimental Study of Effects of Distance, Slope and Break on Putting Performance for Active Golfers." *Science and Golf IV*, 2002.

Cohen, G. L. "On a Theorem of G. H. Hardy Concerning Golf." *The Mathematical Gazette*, 86, no. 505 (March 2002), pp. 120–24.

Coop, R. H. "Mathemagenic Behaviours of Golfers: Historical and Contemporary Perspectives." *Science and Golf III*, 1998.

Cotton, C., and J. Price. "The Hot Hand, Competitive Experience, and Performance Differences by Gender." Munich Personal Re PEc Archive Paper, 2007.

Cross, Rod, and Alan Nathan. "Experimental Study of the Gear Effect in Ball Collisions." *American Journal of Physics*, 75, no. 7 (July 2007), pp. 658–64.

References

Erlichson, Herman. "Maximum Projectile Range with Drag and Lift, with Particular Application to Golf." *American Journal of Physics*, 51, no. 4 (April 1983), pp. 357–61.

Fearing, D., J. Acimovic, and S. Graves. "How to Catch a Tiger: Understanding Putting Performance on the PGA Tour." MIT Sloan Research Paper no. 4768-10, Jan. 18, 2010.

Floyd, Raymond. "Dialogue on Golf." *Golf Digest*, February 1994, pp. 95–109.

Gualtieri, M., T. Tokieda, L. Advis-Gaete, B. Carry, E. Reffet, and C. Guthmann. "Golfer's Dilemma." *American Journal of Physics* 74, no. 6 (June 2006), pp. 497–501.

Hale, T., and G. T. Hale. "Lies, Damned Lies and Statistics in Golf." *Science and Golf*, 1990.

Hardy, G. H. "A Mathematical Theorem about Golf." *The Mathematical Gazette*, 29, no. 287 (Dec. 1945), pp. 226–27.

Hoadley, B. "How to Lower Your Putting Score Without Improving." *Science and Golf II*, 1994.

Johnson, Sal, and Alan Shipnuck. "The Real Cut Streak." *Sports Illustrated*, May 12, 2003.

Jones, R. E., "A Correlation Analysis of the Professional Golf Association Statistical Rankings for 1988." *Science and Golf*, 1990.

Knuth, Dean. "A Two Parameter Golf Course Rating System." *Science and Golf*, 1990.

Kupper, Lawrence, Leonard Hearne, Sandra Martin, and Jeffrey Griffin. "Is the USGA Golf Handicap System Equitable?" *Chance* 14, no. 1 (2001), pp. 30–35.

Libii, Josue Njock. "Dimples and Drag: Experimental Demonstration of the Aerodynamics of Golf Balls." *American Journal of Physics*, 75, no. 8 (Aug. 2007), pp. 764–67.

Lieberman, B. B. "The Effect of Impact Conditions on Golf Ball Spin-rate." *Science and Golf*, 1990.

Lieberman, B. B. "Estimating Lift and Drag Coefficients from Golf Ball Trajectories." *Science and Golf*, 1990.

MacDonald, William, and Stephen Hanzely. "The Physics of the Drive in Golf." *American Journal of Physics*, 59, no. 3 (March 1991), pp. 213–18.

References

McBeath, Michael, Alan Nathan, Terry Bahill, and David Baldwin. "Paradoxical Pop-ups: Why Are They Difficult to Catch?" *American Journal of Physics*, 76, no. 8 (Aug. 2008), pp. 723–29.

Minton, Roland. "A Mathematical Rating System." UMAP Module 725, *UMAP Journal*, 13, no. 4 1992, pp. 317–34.

Minton, Roland. "G. H. Hardy's Golfing Adventure." *Mathematics and Sports*, Mathematical Association of America, 2010.

Minton, Roland. "Lipping Out and Laying Up." *Math Horizons*, Mathematical Association of America, April 2010.

Minton, Roland. "Tigermetrics." *Mathematics and Sports*, Mathematical Association of America, 2010.

Miura, K., and F. Sato. "The Initial Trajectory Plane after Ball Impact." *Science and Golf III*, 1998.

Mizota, T., T. Naruo, H. Simozono, M. Zdravkovich, and F. Sato. "3-Dimensional Trajectory Analysis of Golf Balls." *Science and Golf IV*, 2002.

Mosteller, Frederick. "Lessons from Sports Statistics." *The American Statistician*, 51, no. 4 (Nov. 1997), pp. 305–310.

Murrell, Hugh. "A Mathematical Golf Swing." *The Mathematica Journal*, 1993, pp. 62–68.

Newport, John Paul. "A Stat Is Born: Golf's New Way to Measure Putting." *Wall Street Journal*, March 12, 2010.

Pelz, Dave. "A Study of Golfers' Abilities to Read Greens." *Science of Golf II*, 1994.

Riccio, L. J. "Statistical Analysis of the Average Golfer." *Science and Golf*, 1990.

Rojas, Raul, and Mark Simon. "Like a Rolling Ball." Freie Universitat Berlin, Institut fur Informatik.

Rotella, Bob. "How to Drain Them Like Jack." *Golf Digest*, June 2001.

Scheid, F. J. "On the Normality and Independence of Golf Scores, with Various Applications." *Science and Golf*, 1990.

Scheid, Francis. "A General Principle in Golf." *Science and Golf IV*, 2002.

Scheid, Francis, and Lyle Calvin. "Adjusting Golf Handicaps for the Difficulty of the Course." *Anthology of Statistics in Sports*, 2005.

Smits, A. J., and D. R. Smith. "A New Aerodynamic Model of a Golf Ball in Flight." *Science and Golf II*, 1994.

Stroud, R. C., and L. J. Riccio. "Mathematical Underpinnings of the Slope Handicap System." *Science and Golf*, 1990.

Stroud, Richard, "Proposed Handicap Study." Letter to Gordon Ewen, Chairman, USGA Handicap Procedures Committee, May 1976.

Tavares, G., K. Shannon, and T. Melvin. "Golf Ball Spin Decay Model Based on Radar Measurements." *Science and Golf III*, 1998.

Tierney, D. E. and R. H. Coop. "A Bivariate Probability Model for Putting Proficiency." *Science and Golf III*, 1998.

Turner, A. B. and N. J. Hills. "A Three-Link Mathematical Model of the Golf Swing." *Science and Golf III*, 1998.

Tutelman, Dave. "The Great Square Groove Controversy." Clubmakers' Resource, 1998.

Zagarola, M. V., B. Lieberman, and A. J. Smits. "An Indoor Testing Range to Measure the Aerodynamic Performance of Golf Balls. *Science and Golf II*, 1994.

Books

Albert, Jim, Jay Bennett, and James Cochran, ed. *Anthology of Statistics in Sports*. American Statistical Association, 2005.

Andrisani, John. *The Bobby Jones Way: How to Apply the Swing Secrets of Golf's All-Time Power-Control Player to Your Own Game*. HarperCollins, 2002.

Armenti, Angelo, ed. *The Physics of Sports*. AIP Press, 1992.

Ayres, Ian. *Super Crunchers: Why Thinking-by-Numbers Is the New Way to Be Smart*. Bantam Books, 2007.

Bardi, Jason. *The Calculus Wars: Newton, Leibniz, and the Greatest Mathematical Clash of All Time*. Thunder's Mouth Press, 2006.

Berra, Yogi. *The Yogi Book: I Really Didn't Say Everything I Said*. Workman Publishing Company, 1998.

Blythe, Will. *To Hate Like This Is to Be Happy Forever: A Thoroughly Obsessive, Intermittently Uplifting, and Occasionally Unbiased Account of the Duke–North Carolina Basketball Rivalry*. HarperCollins, 2006.

Bohn, Michael. *Money Golf: 600 Years of Bettin' on Birdies*. Potomac Books, 2007.

Brancazio, Peter. *Sport Science: Physical Laws and Optimum Performance*. Simon and Schuster, 1984.

Cochran, Alastair J., ed. *Science and Golf: Proceedings of the First World Scientific Congress of Golf*. Spon Press, 1990.

Cochran, Alastair J., and Martin R. Farrally, eds. *Science and Golf II: Proceedings of the 1994 World Scientific Congress of Golf*. Taylor & Francis, 1994.

References

Concannon, Dale. *The Ryder Cup: Seven Decades of Golfing Glory, Drama and Controversy*. Aurum Press, 2001.

Cook, Kevin. *Tommy's Honor: The Story of Old Tom Morris and Young Tom Morris, Golf's Founding Father and Son*. Gotham, 2008.

Cook, Kevin. *Driven: Teen Phenoms, Mad Parents, Swing Science, and the Future of Golf*. Gotham, 2008.

Cook, Kevin. *Titanic Thompson: The Man Who Bet on Everything*. Norton and Company, 2010.

Crenshaw, Ben, with Melanie Hauser. *A Feel for the Game: To Brookline and Back*. Doubleday, 2001.

Cupp, Bob. *The Edict: A Novel from the Beginnings of Golf*. Alfred A. Knopf, 2007.

DiPerna, Paula, and Vikki Keller. *Oakhurst: The Birth and Rebirth of America's First Golf Course*. Walker and Company, 2007.

Eastaway, Rob, and John Haigh. *How to Take a Penalty: The Hidden Mathematics of Sport*. Robson Book Ltd., 2005.

Farrally, Martin R., and Alastair J. Cochran, eds. *Science and Golf III: Proceedings of the 1998 World Scientific Congress of Golf*. Human Kinetics, 1999.

Feinstein, John. *A Good Walk Spoiled: Days and Nights on the PGA Tour*. Little Brown and Company, 1995.

Frost, Mark. *The Greatest Game Ever Played: Harry Vardon, Francis Ouimet, and the Birth of Modern Golf*. Hyperion, 2002.

Frost, Mark. *The Grand Slam: Bobby Jones, America, and the Story of Golf*. Hyperion, 2005.

Frost, Mark. *The Match: The Day the Game of Golf Changed Forever*. Hyperion, 2007.

Gallian, Joseph A., ed. *Mathematics and Sports*. Mathematical Association of America, 2010.

Gray, Scott. *The Mind of Bill James: How a Complete Outsider Changed Baseball*. Doubleday, 2006.

Gummer, Scott. *Homer Kelley's Golfing Machine: The Curious Quest That Solved Golf*. Gotham, 2009.

Hardy, G. H. *A Mathematician's Apology*. Cambridge University Press, 1940.

Hoggard, Rex. *The Golf Geek's Bible: All the Facts and Stats You'll Ever Need*. MQ Publications, 2006.

James, Bill. *Baseball Abstract 1984*. Ballantine Books, 1984.

References

James, Bill. *The New Bill James Historical Baseball Abstract*. Free Press, 2001.

Jenkins, Dan. *Dead Solid Perfect*. Main Street Books, 2000.

Jones, Robert Trent, Jr. *Golf by Design: How to Lower Your Score by Reading the Features of a Course*. Little, Brown and Company, 2005.

Jorgensen, Theodore P. *The Physics of Golf*. Springer-Verlag, 1994.

Kelley, Homer. *The Golfing Machine: Geometric Golf: The Computer Age Approach to Golfing Perfection*. Golfing Machine Press, 2006.

Lennard, Duncan. *eXtreme golf: The World's Most Unusual, Fantastic and Bizarre Courses*. Sourcebooks, Inc., 2004.

Lewis, Chris. *The Scorecard Always Lies: A Year Behind the Scenes on the PGA Tour*. Simon & Schuster, 2007.

Littlewood, J. E., and B. Bollobás. *Littlewood's Miscellany*. Cambridge University Press, 1986.

Lorenz, Edward N. *The Essence of Chaos*. University of Washington Press, 1993.

Macrury, Downs. *Golfers on Golf: Witty, Colorful, and Profound Quotations on the Game of Golf*. Signature Press, 2007.

Mediate, Rocco, and John Feinstein. *Are You Kidding Me? The Story of Rocco Mediate's Extraordinary Battle with Tiger Woods at the U.S. Open*. Little Brown and Company, 2009.

Montville, Leigh. *The Mysterious Montague: A True Tale of Hollywood, Golf, and Armed Robbery*. Doubleday, 2008.

Murphy, Michael. *Golf in the Kingdom*, Viking Press, 1972.

Neimark, J. I., and N. A. Fufaev. *Dynamics of Nonholonomic Systems*. American Mathematical Society, 1972.

O'Connor, Ian. *Arnie and Jack: Palmer, Nicklaus, and Golf's Greatest Rivalry*. Houghton Mifflin, 2008.

O'Connor, Tim. *The Feeling of Greatness: The Moe Norman Story*. Master Press, 1995.

O'Keefe, Michael, and Teri Thompson. *The Card: Collectors, Con Men and the True Story of History's Most Desired Baseball Card*. William Morrow, 2007.

Pelz, Dave, with Nick Mastroni. *Putt Like the Pros: Dave Pelz's Scientific Way to Improving Your Stroke, Reading Greens, and Lowering Your Score*. Harper & Row, 1989.

References

Rodel, Chris, and Allan Zullo, *Amazing But True Golf Facts: 2008 Calendar.* Andrews McMeel Publishing, 2008.

Russell, Gordon. *Sport Science Secrets From Myth to Facts.* Trafford Publishing, 2001.

Sampson, Curt. *The Eternal Summer: Palmer, Nicklaus, and Hogan in 1960, Golf's Golden Year.* Villard Books, 1992.

Schrier, Eric, and William Allman, eds. *Newton at the Bat: The Science in Sports.* Macmillan, 1987.

Schwarz, Alan. *The Numbers Game: Baseball's Lifelong Fascination with Statistics.* Thomas Dunne Books, 2004.

Smith, Robert, and Roland Minton. *Calculus: Early Transcendental Functions.* 3rd ed. McGraw-Hill, 2007.

Stanton, Tom. *Ty and the Babe: Baseball's Fiercest Rivals.* Thomas Dunne Books, 2007.

Thain, Eric, ed. *Science and Golf IV: Proceedings of the World Scientific Congress of Golf.* Routledge, 2002.

Torrey, Lee. *Stretching the Limits: Breakthroughs in Sports Science That Create Super Athletes.* Dodd, Mead and Company, 1985.

Wade, Don. *Talking on Tour: The Best Anecdotes from Golf's Master Storyteller.* McGraw-Hill, 2001.

Werner, Frank, and Richard Greig. *How Golf Clubs Really Work and How to Optimize Their Design.* Origin Inc., 2000.

Werner, Frank, and Richard Greig. *Better Golf from New Research.* Origin Inc., 2001.

Wishon, Tom, with Tom Grundner. *The Search for the Perfect Golf Club.* Sports Media Group, 2005.

Zumerchik, John. *Newton on the Tee: A Good Walk through the Science of Golf.* Simon and Schuster, 2002.